DEEP-SKY OBJECTS FOR BINOCULARS

DEEP-SKY OBJECTS FOR BINOCULARS

JOHN T. KOZAK

Sky Publishing Corporation
Cambridge, Massachusetts

1988

49 Bay State Road
Cambridge, Mass. 02138

ISBN 0-933346-50-2

CONTENTS

PREFACE

It's good to see more observers using binoculars these days. At last, these underrated and often ignored instruments are receiving some of the respect and attention they deserve. For too long the binocular has suffered from a reputation as a secondary instrument for the serious amateur. Now more observers are discovering what a few of us old binocular jockeys have known all along: these are truly first-class instruments in their own right.

Comet Halley deserves some of the credit. Many observers reported that binoculars, rather than large telescopes, gave the best view of the great comet. However, most of the credit for this revival should go to the authors of books and magazine articles on binocular astronomy. Books by Leslie Peltier, James Muirden, and Patrick Moore have brought the fundamentals of amateur astronomy, binocular style, to thousands of new observers. Just as noteworthy are the many excellent articles by James Haas, David Williams, Craig Crossen, and George Lovi for *Astronomy* and *Sky & Telescope* magazines. These authors have done binocular astronomy a real service by demonstrating that the binocular can indeed be used for serious work.

This book is designed to complement the works of these authors. It will take the amateur beyond the introductory level and into the fascinating area of deep-sky observing. I hope it will also prompt others to contribute to this small but growing collection of works on binocular astronomy. Much remains to be written.

INTRODUCTION

The expression "deep-sky object" is an amateur's term for any object outside the solar system that appears as something other than a single star in a binocular or telescope. Although some consider double stars to be deep-sky objects, many authors (including this one) tend to treat them separately and prefer to reserve the term "deep-sky object" for star clusters, nebulae, and galaxies. Thus, double stars will not be treated in this book.

Literally hundreds of deep-sky objects are visible in a binocular, even if you exclude double stars. This book contains descriptions of over 130 clusters, nebulae, and galaxies found in northern skies. Some are visible only in an 11 x 80 "giant," but dozens are visible with nothing more than an 8 x 30 "mini." Anyone who has access to a binocular can observe and enjoy these fascinating objects.

This is a manual for the intermediate-level amateur who has some grasp of descriptive astronomy and some familiarity with the night sky. It is a book for the amateur who wishes to go a step beyond the introductory level, but who may not have access to a telescope. Anyone who has priced even a small telescope will understand why there are so many such amateurs today.

Deep-Sky Objects for Binoculars is divided into three chapters. These will take an observer from the nuts and bolts of binocular design to the actual observing of deep-sky objects in short order.

Chapter 1 is all equipment — the essentials of binocular design and quality. Included here is a complete rundown on binocular basics as well as recommendations concerning magnification and aperture for various deep-sky objects. There are also recommendations concerning brands I have personally field tested. Be sure to read this chapter before buying a binocular.

Equipment, however, is only a small part of observing. It takes well-developed observing skills to detect many deep-sky objects with a binocular. Chapter 2 outlines the necessary attitudes and techniques, and offers some helpful tips on selecting an observing site. This chapter is required reading for any new observer.

Chapter 3 contains brief descriptions of over 130 deep-sky objects as they appear in both standard and giant binoculars. This list is hardly the limit, but it does make for a solid observing program. The objects range in difficulty from very easy to rarely possible and include examples of each type of deep-sky object.

This book is designed to be used — a fireside reference it is not. It provides a solid base for the new observer, but not so much detail as to overwhelm and intimidate. This is more a book on how to observe and which objects to observe than a collection of personal observing notes. Make your own notes, and describe what you see in your own words. That's half the fun of observing.

Dark skies to all!

DEEP-SKY OBJECTS FOR BINOCULARS

Chapter 1

DEEP-SKY BINOCULARS

The Parable of Observer A and Observer B

Most people have a natural curiosity about the stars and heavens, yet relatively few select astronomy as a hobby. Those who do may be rewarded with a pursuit lasting a lifetime, but all too often a new observer abandons his or her hobby after an initial romance with the sky. Why?

Consider the plight of one enthusiast. Let's call him Observer A. After securing a loan, Observer A eagerly purchases a new "Superstar Quantum X" telescope and the "expert" grade accessory package, which includes enhanced optics, two extra-wide-angle eyepieces with illuminated reticles (and rotary eyepiece holder to hold them), a set of three Barlow lenses, erecting prism, star diagonal, a complete selection of solar and nebular filters, jumbo finder scope, piggyback guide scope, 2" adapter assembly, ball-bearing equatorial mount with extra counterweights, multi-functional digital dual-axis quartz drive corrector with AC adapter, manual declination controls, setting circles, tele-compressor, tele-extender, camera adapter, dew zapper, mahogany tripod, carrying case, and, of course, much, much more.

Unfortunately, Observer A must also purchase one of those large backyard storage sheds, since a Superstar

Observer A.

Quantum X and expert accessory package won't fit in a closet or even a living room. A Superstar Quantum X won't fit in a compact car either, so Observer A rounds out his equipment package with a used cargo van.

With great expectations (and even greater equipment payments), Observer A proceeds with his observing program. He now learns that equipment costs are measured in more than just dollars. It takes the better part of an hour to run down his equipment checklist and load the van with telescope and accessories. It takes even longer to set up the equipment at his observing site. And of course it takes a like amount of time to pack up and go home. At this point, Observer A is not sure if his sessions are observing sessions or equipment sessions.

Worse yet, the equipment does not perform to his expectations. Because the polar alignment is never quite right, his setting circles and drive are always a little off. And it seems that something is forever working itself loose and demanding attention. All he does is adjust equipment. And talk about embarrassing! When a friend asks to see the Andromeda galaxy, he has to tell her to come back after he learns how to use all his new equipment.

But the biggest disappointment of all for Observer A is at the eyepiece of his telescope. Why don't those objects look like the pictures in the magazines and telescope brochures? After all, 600x should be enough magnification for any object. How could a Superstar Quantum X let him down? He bought all the right accessories! Maybe if he bought a bigger telescope. . . .

His ''for sale'' ad appears in the next issue of his favorite astronomy magazine. With luck, he'll make enough on his old telescope to cover the down payment on a Superstar Quantum Double X Plus — the new computer-controlled model that can turn a backyard into a professional observatory. Of course, it may take some time to sell his

Observer B.

old telescope, so in the meantime Observer A decides to take up golf.

Observer B is from a lower economic class. He is thoroughly crushed when his telescope loan is rejected. Almost. Rather than give up on astronomy, he seeks an affordable alternative. He finds it in the form of an old 7 x 50 binocular at a neighbor's garage sale. It's not as glamorous as a telescope, but it should be good for a few deep-sky objects. To round out his equipment package, he gets a star map and covers the lens of a flashlight with a piece of red plastic from an old notebook divider. With the addition of a wire-bound notebook and a pencil, and a sack to carry these few items, Observer B is now ready to observe.

Observer B also finds that his equipment does not perform to his expectations. It's much better. There's nothing to set up and adjust. Instead of a few interesting celestial objects, he finds dozens and dozens. There always seems to be something more to observe with that old 7 x 50. After a few months his deep-sky list grows to 50 objects. But he's just getting warmed up. Wait till he takes that old 7 x 50 on his next camping trip. Look out 100 objects!

But what Observer B likes best about the old 7 x 50 is the "hands on" way he can observe deep-sky objects. There's a bare minimum of equipment between him and the wonders of the night sky. He finds it extremely satisfying to locate deep-sky objects without resorting to setting circles or a computer. And now, with his newfound knowledge of the night sky, Observer B has company when he observes. (There's a girl in every story.)

This little parable is not a bit overdone. You really can observe scores of clusters, nebulae, and galaxies and learn the fundamentals of observing with nothing more than an old 7 x 50. Actually, many deep-sky observers use a binocular side by side with their telescopes. They know that the binocular is more convenient and more effective than a telescope at times. In fact, the binocular is so easy to use that some deep-sky experts such as George Lovi, in *Sky & Telescope* magazine, recommend the binocular as a first instrument for all newcomers to astronomy.

The cost difference between the binocular and the telescope is also no exaggeration. It's hard to believe, but you can pay more for a single premium-grade telescope eyepiece than a good, serviceable 7 x 50 binocular. Amateur astronomy can be an expensive hobby, but it doesn't have to be. If you're putting off deep-sky observing until you can afford expensive equipment, you're missing a valuable opportunity to enjoy the night sky and develop basic observing skills. You don't need a telescope to do this. All you need is a binocular.

Just ask Observer B.

Binocular Basics

Binoculars come in a bewildering array of sizes, designs, and options. There are models intended for birdwatching, hunting, sportswatching, military maneuvers, marine use, and yes, even astronomy. Thus it might be tempting to assume that it takes a very specialized binocular to observe deep-sky objects. This is absolutely wrong. Even a common 7 x 35 will reveal dozens of them. In fact, your present binocular may be all you need to begin your search for clusters, nebulae, and galaxies.

Of course, some binoculars are much better for this purpose than others. The size and type of binocular you use affects the number of objects you can see, the amount of detail you can see in them, the type of support your instrument will need, and, of course, the price you will have to pay. The kind of instrument may even affect the type of deep-sky object you observe, since no one binocular is best for all the different types.

To determine the suitability of any binocular for deep-sky observing, you will first have to learn some binocular basics and then balance all the variables against your needs and preferences. First the basics:

Magnification. The first number on the label of a binocular, for instance, the 7 in 7 x 35, is the magnification. Standard-size binoculars range between 6x and 10x. "Giants" can run as high as 30x. Zoom models, which allow the magnification to be varied mechanically or electronically, rarely have the optical quality needed for deep-sky work.

Aperture. The second number on the label is the aperture or width of the objective (front) lens in millimeters. Standard-size binoculars have apertures of 35 to 50 millimeters. Giants typically have apertures between 60 and 80 millimeters and may range as high as 150 millimeters. Aperture is the name of the game in deep-sky work since it determines the ability to detect faint objects. Binoculars with apertures of less than 35 millimeters do not gather enough light to reveal large numbers of deep-sky objects.

Field of view: This is the amount of sky or land that can be seen when you look through a binocular. It can be expressed in degrees, or as feet at 1,000 yards (52 feet at 1,000 yards equals 1°). Wide field is perhaps the major advantage of the binocular over a telescope. It allows an observer to scan large areas of the sky quickly and easily. The field of view decreases as magnification increases, but binoculars of 6x to 10x have fields large enough for any deep-sky object.

Eye relief. This is the distance between the eyepiece of the binocular and the point where your eye must be placed to see the full field of view.

If you wear eyeglasses for astigmatism, be sure to choose a long eye-relief binocular. In a short eye-relief model you won't be able to position your eyes close enough to the eyepieces to enjoy the full field of view unless you remove your glasses. To check for possible field loss, simply compare the size of the field with and without glasses. A small loss of field with glasses on is acceptable, but a large loss (as much as 50 percent in some models) is not. Eyeglass wearers who suffer only from near- or far-sightedness can observe without glasses and not lose any performance since these defects can be corrected by simply focusing the binocular. Astigmatism cannot be corrected by focusing.

Coatings. These are chemical deposits (metallic salts) on the lens and prism surfaces that dramatically increase light transmission. Since light transmission is so critical for detecting faint objects, you should pay particular attention to the type of coatings on a binocular. Several trade terms are used in reference to coatings.

"Coated" binoculars will have at least one air-to-glass surface coated with magnesium fluoride. A properly applied coating of magnesium fluoride will reduce light loss at an air-to-glass surface from about 4 percent (this is the average light loss for an untreated surface) down to 1.5 percent.

"Fully coated" binoculars will have all air-to-glass surfaces coated with magnesium fluoride. This is a significant improvement over an untreated model (which may have a total light loss of 50 percent), since there are many air-to-glass surfaces in a binocular. A good fully coated binocular will have a light loss of only 15 percent.

This should be considered a minimum for a deep-sky binocular.

"Multi-coated" binoculars have a more efficient coating than standard magnesium fluoride. As the term indicates, multi-coating is a process that deposits several layers of chemicals on an optical surface rather than a single layer. Every manufacturer has its own multi-coating formula that it markets under its own trade name. Because of these different formulas, a multi-coated lens may range in color from green to violet. However, magnesium fluoride coatings may also appear reddish-purple or even green if incorrectly applied, so it is unwise to rely on lens color to determine the type of coating on your binocular. If in doubt, check with the manufacturer.

A multi-coated surface reduces light loss from 4 percent for an untreated surface down to 0.5 percent or so. Compare this to the 1.5 percent for a magnesium fluoride surface. Eventually multi-coating will replace single-layer magnesium fluoride coating as a standard lens coating, but for now you should be aware of the difference.

Fully multi-coated binoculars have all air-to-glass surfaces multi-coated. On the average this type of model will cost $50 to $75 more than a magnesium fluoride coated model, but it is well worth the difference in price since it can mean an increase of 15 or 20 percent in performance. A good fully multi-coated model will have a light loss of only 5 percent. This is the ultimate for a deep-sky binocular.

Focusing. Two common systems are used to focus binoculars. Both are suitable for astronomy as long as they render sharp images that remain in focus. (Nothing is more aggravating than a binocular that won't stay in focus.)

Binoculars with an individual-focusing system require

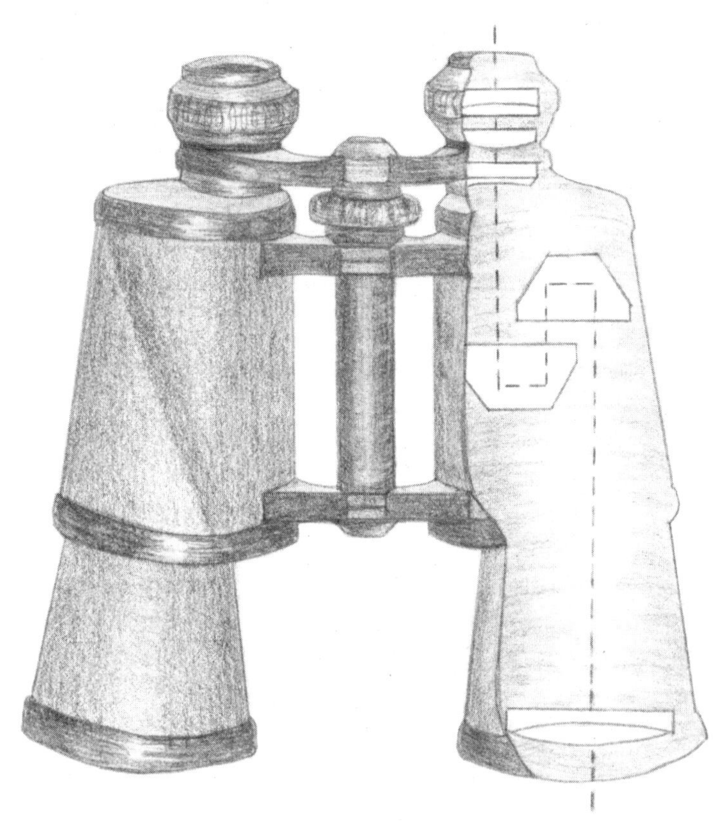

Light path through a Porro-prism binocular.

each eyepiece to be turned independently. This system is tedious and time-consuming when more than one person is using the binocular, or when you refocus often for nearby objects as in birdwatching. But it is the easiest system to waterproof, fogproof, and dustproof, and it is exceedingly durable. For this reason, most military binoculars have individual focus. The newest version of this style allows you to focus each side once for all objects from 20 yards to infinity. In my opinion, however, this design does not focus as sharply and produces more eyestrain than the older style of individual focus.

Binoculars with a center-focusing system, on the other hand, allow both eyepieces to be focused simultaneously by means of a central wheel or lever. Usually the right eyepiece can also be focused separately to compensate for differences in strength between the right and left eye. Because this system is quicker than an individual-focus system, it produces less eyestrain and is generally more convenient in most applications. But it should not be considered an advantage for astronomy, where you will set the focus to infinity and leave it there.

Prism Type. The prisms in a binocular right the upside-down images produced by the objective (front) lenses and also ensure correct right-left orientation. Modern binoculars use one of two prism systems.

''Porro-prism'' binoculars have the traditional offset or dogleg shape with the front lenses not in line with the eyepieces. (See facing page.) The older German or ''Z'' style Porro-prism binocular has a three-piece body with lens tubes that can be unscrewed (by a qualified repairperson) from the mainframe. The newer American or ''B'' style Porro-prism binocular has an improved one-piece body that is considerably more durable and less likely to get knocked out of alignment.

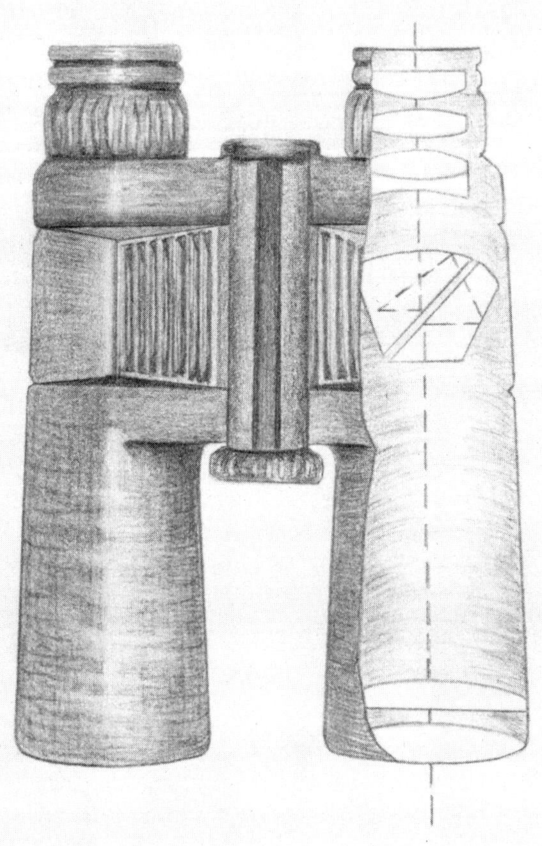

Light path through a roof-prism binocular.

The Porro prisms themselves can be made of two types of glass. Prisms made of BK-7 glass are cheaper but less light-efficient than prisms made of the more desirable BAK-4 glass. Binoculars that combine BAK-4 prisms and fully multi-coated optics are the most light-efficient yet made. Hold the binoculars about a foot in front of you up to a bright sky. If the little disks of light floating behind the eyepieces have square shadowing in their edges, the prisms are the cheap kind.

''Roof-prism'' binoculars have a straight tube shape with eyepieces located directly behind the objectives. (See facing page.) This makes for a trimmer and lighter binocular, but one that is slightly less light-efficient due to some light loss at a mirror surface that is part of the roof prism. In addition, prism alignment is far more critical. A roof-prism system can have an angular error in its alignment of no more than two seconds of arc (much less than the width of a human hair) or the image will suffer. Compare this to the huge 600 seconds of arc error that is allowable in a Porro-prism system. This explains why Porro-prism binoculars are generally cheaper than roof-prism binoculars and why it is rare to find roof-prism binoculars selling for less than $100.

Nevertheless, a good roof-prism binocular is an efficient and very comfortable instrument for deep-sky observing. Its lighter weight and slimmer profile offset the negligible additional light loss inherent in its design, especially during long observing sessions. For deep-sky work, aperture, coating type, and lens quality are more important than prism type.

Exit pupil. This is the beam of light that passes or ''exits'' from the eyepiece of the binocular. It can actually be observed as the small circle of light behind the eyepiece when the binocular is held about a foot away from the eye. The exit pupil is measured in millimeters and can be easily calculated by dividing aperture by magnification. Thus a 7 x 35 has an exit pupil of 5 mm. A 7 x 42 has an exit pupil of 6 mm. Large exit pupils are very desirable for deep-sky observing because they provide the highest possible surface brightnesses for very faint objects. A galaxy that is too faint to detect in a 10 x 40 (4-mm exit pupil) often becomes obvious in a 10 x 70 (7-mm exit pupil). For this reason, binoculars with exit pupils of 5 mm or more are best for deep-sky observing. Binoculars with exit pupils of more than 7 mm are not practical, however, since the human eye does not open wide enough to accept an exit pupil larger than 7 mm.

Relative brightness or luminosity factor. This expresses a mathematical relationship between binoculars with different exit pupils. It is calculated by squaring the exit pupil diameter in millimeters. A binocular with an exit pupil of 3 mm has a relative brightness of 9, while one with an exit pupil of 6 mm has a relative brightness of 36. This tells you that doubling the exit pupil results in more than a doubling of image surface brightness — it results in a fourfold increase.

Twilight factor. This is a measure of how much detail can be seen in low-light conditions. It is the square root of the magnification multiplied by the objective diameter in millimeters. At first glance this might seem to be a good way to assess the potential of a binocular for deep-sky work. Unfortunately, it overemphasizes the role of magnification. Twilight factor is a more useful index for terrestrial applications.

Performance. It is important to make a distinction between a binocular's potential and its performance.

Potential is purely theoretical and mathematical, and is expressed as relative brightness or twilight factor. It says nothing about optical quality, precision of manufacture, or viewing comfort. These are part of a binocular's performance. Thus a quality 10 x 40 may outperform a cheap 10 x 50, even though its theoretical potential is less.

Accessories. Unlike the typical telescope, a binocular needs few accessories. Some, however, are essential for deep-sky observing.

Tripods. These are essential to steady binoculars of over 15x, and can save wear and tear on neck and arms with large binoculars of 10x to 15x. If you buy a giant, be sure it has a tripod adapter.

Tripods have other uses as well. They allow you to share deep-sky objects with new observers or interested bystanders. For this reason I usually carry one in the trunk of my car. Never pass up a chance to make a convert to astronomy! Tripods also free up both hands for sketching, note-taking, and other activities.

Unfortunately, tripods are very difficult to use for objects that are high overhead. You can always crawl underneath a tripod, but it is usually far easier to use a mirror-type binocular holder.

Binocular holder. This device allows an observer to scan overhead constellations comfortably without looking up. The binocular is held in a mount that points it downward to a mirror that reflects overhead objects. Bryan Shumaker's design is simple, cheap, and effective. (See *Astronomy* for February, 1988.)

Filters. These are not commonly sold for binoculars, but can be obtained directly from manufacturers. Specialized "nebula" or "light-pollution" filters are useful for detecting elusive nebulae.

Star maps. A star map is, of course, essential for locating

Basic equipment of the binocular observer. Shown is a 7 x 42 Brunton roof-prism binocular — a versatile and affordable fully coated model used to collect much of the data for this book.

almost all deep-sky objects. Since most objects are found by referring to naked-eye stars, a 6.5-magnitude map is satisfactory as long as it shows a sufficient number of deep-sky objects. More detailed maps, however, offer definite advantages. (See Appendix II.)

Red flashlight. Red light is the least disruptive to dark-adapted eyes. You can buy a red-lens flashlight to read your star map, or you can make your own by covering the lens of an ordinary flashlight with red plastic, red construction paper, or by painting the bulb red with fingernail polish.

Notebook. Always take notes. Don't trust your memory. All writing should be in pencil, since pencil lead won't run when wet.

Straps. Binoculars come with straps. Use them. Binoculars rarely wear out. Most are damaged when dropped or swung against hard objects.

Recommendations for Deep-Sky Binoculars

Magnification

Compared to telescopes, all binoculars are low-powered instruments. Thus it might be tempting to assume that there is no significant difference between a binocular of 7x and one of 15x. This is definitely not so. Even a small change in magnification can affect the way a binocular is used and alter its suitability for observing different types of deep-sky objects.

Magnification, of course, is desirable. But it does have its drawbacks. As magnification increases, image surface brightness decreases. This presents no problem when a bright image is magnified, but when a very faint object is magnified it may disappear entirely. The only way to avoid this loss of image brightness due to magnification is to increase aperture also. Unfortunately, more aperture means more size, weight, and expense. For instance, a pair of top-quality 7 x 42 binoculars may weigh about 2 pounds and cost $500; but a 25 x 150 with the same 6-mm exit pupil will weigh almost 70 pounds and cost nearly $9,000.

Field of view also decreases when magnification is increased. In practice this is only a problem for binoculars of over 20x, since lower-powered instruments have ample fields for most deep-sky objects.

The single biggest drawback of magnification becomes apparent when you attempt to steady a high-magnification binocular. Most people can hold a 7x or 8x model reasonably still without support, but a 10x or 11x model will usually need some type of brace — a car roof, tree limb, almost anything. Binoculars of over 15x require something more substantial, such as a tripod or even a telescope mount, both to steady them and to support their greater weight.

The primary use of magnification in a deep-sky binocular is to show detail in large objects and to separate small ones from surrounding stars. As a very general rule of thumb, objects listed as being less than 5′ in size will appear as little more than a point of light or "fuzzy" star at magnifications of less than 15x. (Published sizes for deep-sky objects are usually based on long-exposure photographs that invariably record more size and detail than can be seen with amateur telescopes and binoculars.) To distinguish these small deep-sky objects from stars, you will need a good star map and top-notch map reading skills, as well as some experience as an observer.

Choose the magnification of your deep-sky binocular with care. Too much power is far worse than too little, especially if it decreases the exit pupil below 5 mm or makes your binocular awkward to handle. With this in mind the following recommendations can be made.

7x, 8x Binoculars

These are the easiest and most comfortable magnifications to use for deep-sky observing. Most 7x or 8x binoculars are small and light enough to be held by hand for long periods of time without tiring. In addition, their wide fields make it easy to locate objects and navigate among the stars.

These are best for the largest deep-sky objects such as widely scattered open clusters and the larger diffuse nebulae. (They're also great for comet watching.) They are unmatched for leisurely scanning the Milky Way on a dark night.

Binoculars of 7x or 8x will also pick up most of the globular clusters that are bright enough to be seen in a binocular, and will show the outline of many nebulae and even some of the larger galaxies. They are not, however, as effective as higher-magnification models for spotting small galaxies or resolving detail in star clusters.

9x, 10x, 11x Binoculars

These can be hand held for short periods (grasp them up front near the objectives) but will need some type of support for extended viewing. This need not be a tripod, as some authorities insist. I have had excellent success with an 11 x 80 supported with elbows resting on the arms of a lawn chair or the roof of a car. Tripods have their uses, but they rob you of much of the freedom of movement that is the joy of observing with binoculars.

Binoculars of 9x to 11x are the best choice for an all-purpose deep-sky binocular. They are small enough to be handled without a tripod, but show significantly more detail and reveal smaller objects than 7x or 8x models, yet have sufficient fields of view for all but the very largest objects. Only for the smallest objects will you need more magnification than this.

Steiner Military/Marine 15 x 80 giant binocular on tripod. Fine galaxy spotter.

15x-30x Binoculars

These always require a tripod. Binoculars of this power are specialty instruments designed to detect small galaxies or planetaries and to resolve star clusters. They are something of a transition between a standard binocular and a small telescope.

I do not recommend these binoculars for a first instrument. Their mounting requirements and narrow fields of view make them difficult to use, and unless they have exit pupils of 5 mm or more, they will not be as effective as smaller instruments for rich-field viewing.

Aperture and Exit Pupil

The ability of a binocular or telescope to reveal faint celestial objects is a function of both objective size (aperture) and magnification. In a telescope this ability to reveal faint objects (or collect light) is discussed in terms of aperture alone, since magnification can be changed easily as needed. Not so in a binocular. Here magnification is fixed and will directly affect image surface brightness and the ability to detect very faint objects. For this reason it is most accurate to assess the potential of a binocular to reveal faint objects by referring to its exit pupil as well as its aperture.

A deep-sky binocular should have an exit pupil of at least 5 mm. An exit pupil of 6 mm or 7 mm is even better. Smaller exit pupils will pick up the brightest deep-sky objects, but are severely limited for wide-field work. My personal preference in a deep-sky binocular is a 7-mm exit pupil.

Some observers, however, prefer smaller exit pupils. They claim that a smaller exit pupil will increase the contrast between faint objects and the sky background, making it easier to spot difficult objects, especially under light-polluted conditions. The smaller exit pupil supposedly doesn't brighten up the sky background as much as a larger exit pupil.

I have to dispute this theory. After testing dozens of binoculars under all types of seeing conditions, I have to report that a smaller exit pupil binocular has never outperformed a larger exit pupil binocular of the same power and comparable quality when observing faint deep-sky objects. I have yet to see a good 10 x 50 reveal more objects or show more detail than a good 10 x 70, even under light-polluted conditions. This is easy enough to test, so I suggest you experiment for yourself.

To be fair, I must admit that not everyone has eyes that can accommodate a 7-mm exit pupil, especially older observers. (The pupil of the eye can shrink with age.) In this case the 7-mm exit pupil is a light waster. Nevertheless, it won't put anyone at a disadvantage except, perhaps, for increased weight and cost.

Limiting Magnitude

The theoretical limiting magnitude (dimmest visible star) for any binocular depends on aperture. (The exit pupil determines the whole field's surface brightness.) Borrowing some data for telescopes, the limiting magnitude for various apertures looks something like this:

25-mm aperture	8.8 magnitude
50-mm aperture	10.3 magnitude
80-mm aperture	11.2 magnitude

There is evidence that using two eyes to observe instead of one can increase the visibility of low-contrast objects. Several researchers claim an increase of up to 40 percent in the perception of low-contrast objects when using binocular vision. If so, you can add 0.4 to the limiting magnitudes above. (For a more detailed discussion, see

Astronomy for November, 1986, and *Sky & Telescope* for June, 1985.)

So much for theory. In practice, these limits are rarely reached. Light pollution, poor seeing conditions, defects and variations in the observer's eyesight, and shortcomings in the optical quality of the binocular can all reduce the limiting magnitude for a given aperture. Only on exceptional nights will your binocular perform at or near its limiting magnitude.

Furthermore, there is tremendous variation in the visibility of deep-sky objects, even those that have the same published magnitude. (Published magnitudes usually treat all objects as if they were single points of light, even though many of them have their light spread over a large area of the sky.)

For these reasons you shouldn't expect to find too many 10th-magnitude galaxies with a 7 x 50 (limiting magnitude 10.3) under ordinary conditions. This is especially true if your site is even slightly light-polluted, as most are. Indeed, light pollution is the single greatest menace to amateur astronomy and is usually the limiting factor on performance for most observers.

If you need more performance, it usually makes more sense to move to a darker site rather than move up to a larger aperture. Not only is this strategy cheaper, it is also more effective. Moving from a 50-mm to an 80-mm binocular (the largest common binocular aperture) results in a gain of only one magnitude, but light pollution can rob you of two or three magnitudes. Thus there is simply no reason for the observer who uses a binocular to come down with a case of "aperture fever."

What kind of performance can you expect in the field for various apertures? As we have seen, this depends on a great many things. Under moderately dark rural skies I average something like this:

Bushnell Explorer II 7 x 50 Porro-prism binocular. A fully coated utility-grade model capable of doing serious deep-sky work.

50-mm objective, 7-mm exit pupil

- 7th-magnitude objects easy
- 8th-magnitude objects may require searching
- 9th-magnitude objects require careful scanning, can bc difficult
- 10th-magnitude objects a challenge, not always possible.

80-mm objective, 7-mm exit pupil

- 7th-, 8th-magnitude objects easy
- 9th-magnitude objects require careful scanning
- 10th-magnitude objects usually difficult, require a good night
- 11th-magnitude objects only on a rare night.

In my opinion, you can usually subtract a little less than one magnitude when you drop to a 5-mm exit pupil with the objectives listed above. You can subtract one to two magnitudes for objects that are within 20° of the horizon. And as mentioned before, you can subtract up to three magnitudes for light pollution.

Remember, these are guidelines based on my experience and my observing sites. The magnitude you squeeze out of your binocular will depend on your experience and your sky quality, as well as the type of object you observe. I can promise you, though, that 50 mm is enough aperture to keep you busy for a long time.

Binocular Quality

It is difficult for a binocular nut like myself to maintain a perspective on binocular quality. It would be easy, but terribly misleading, to claim that only the best quality binoculars are suitable for astronomy. The truth is, much of the data for this book was collected with ordinary department-store binoculars costing about $125. If you own a binocular, go out under the stars and use it! Don't wait until you can afford something better. Besides, you'll be in a better position to appreciate and evaluate a good binocular after you've gained some experience as an observer.

What is a quality binocular? For astronomy, it is first and foremost one that transmits a high percentage of the light it collects. It should also produce sharp, pinpoint star images, preferably out to the edge of the field. Finally, a quality binocular should be mechanically sound and durable enough to provide a lifetime of reasonable use.

All binoculars lose some of the light as it travels through the many lenses and prisms. This light loss can be as high as 50 percent in a cheap, uncoated model or as low as 5 percent in the best multi-coated one.

As a minimum for deep-sky work, a binocular should be fully coated. This means that all air-to-glass surfaces will be coated with magnesium fluoride. Such coating will reduce light loss to about 15 percent in the better models. This is certainly adequate for astronomy, and will produce bright, crisp images if combined with lenses of suitable aperture and quality.

The very best binoculars for astronomy will be fully multi-coated and feature prisms of BAK-4 glass. Binoculars with this combination will have a light loss of only about 5 percent. Moderately priced binoculars of this type are usually so labeled as a selling point. Top-of-the-line binoculars, however, feature the best in coatings and prisms as standard equipment and are rarely labeled with this information.

Properly ground and polished lenses are the heart and soul of a quality binocular. They are also the biggest part of the price tag. In cheap binoculars, stars appear tear-shaped, streaky, or fuzzy due to poor quality lenses. This makes it very difficult to distinguish a small galaxy or planetary from a star. A deep-sky binocular should render crisp, point-like star images, at least at the center of the field. Moderately priced binoculars may be expected to show some distortion at the edge of the field, but top quality binoculars should not.

A quality binocular should also be properly aligned. Even a slightly misaligned pair can produce a double-dose aspirin headache in short order. That's why professionals who use a binocular on a daily basis (guides, wardens, biologists) use only the best.

A well-made binocular should also be mechanically sound and be durable enough to give many years of service. For astronomy a binocular need not be shockproof, waterproof, or rubber armored, unless you plan to take it on an expedition into the mountains or use it on military maneuvers. It should, however, have a dependable focusing system that works quickly, smoothly, and positively. You should not have to constantly refocus a quality binocular.

How much does quality cost? Binoculars that cost less than $100 (1988 prices) are always a gamble. They rarely have the lens quality and coatings needed for serious work. Worse yet, they will give only limited service before major problems such as fogging and misalignment develop. If you buy a binocular in this price range, you may soon find yourself shopping for another.

Binoculars in the $100 to $225 price range will give you the most for your money. These models feature fully coated optics at the low end of this range and some multi-coating at the high end. The optical quality at this price runs from fair to good — certainly adequate for deep-sky work. And at this price, you can expect years of service, even a lifetime of use if you handle your binocular with a little care.

There are many brands and model choices in this price range, but some of my favorites are the Tasco World Class

Zeiss B series 8 x 56 roof-prism binocular. One of the most comfortable models on the market for eyeglass wearers and a superb deep-sky instrument.

series, the Bushnell Explorer series, the Bausch & Lomb Custom series, the Swift Audubon series, and house brand binoculars sold by camera companies such as Nikon and Pentax or by telescope manufacturers such as Celestron and Meade. There are also many dozens of companies who buy binoculars in Japan and put their own labels on them. Some of these are excellent.

At the higher end of this range are the lowest-priced giants, those of standard quality and magnesium fluoride coatings sold by Swift, Meade, Celestron, Orion, and Parks. These giant binoculars are an excellent buy for someone who wants to get serious about deep-sky work at a reasonable price.

Binoculars in the $225 to $400 price range feature multi-coatings and noticeably better optics. Some examples in this price range are standard-size binoculars by Steiner and by aus Jena. The most important binoculars in this price range, however, are giants with multi-coatings and BAK-4 prisms. These are usually referred to as Custom, Enhanced, Deluxe, or Special models. They are the ultimate deep-sky instruments.

Binoculars above $400 are premium quality instruments incorporating the latest in binocular technology and construction. Everyone should have the opportunity to observe with one of these instruments. Stars are unbelievably crisp all the way to the edge of the field. Eyestrain is virtually eliminated. Light transmission for most models is near 95 percent. You'll never have cause to regret the purchase of a Zeiss, Leitz, Swarovsky, or Fujinon unless you lose it or have it stolen. Some models I have tested are the Zeiss Dialyt 7 x 42B and 8 x 56B (my favorite "mini" giant); the Leitz Trinovid 8 x 40; the Bausch & Lomb Elite 8 x 42; the Swarovsky 7 x 42 and 8 x 56 (impressive); and the Steiner Military 15 x 80 (great galaxy spotter). Also excellent by reputation are the Nikon Protostar and the

Fujinon F series. I haven't tested these, but they're on the top of my list to try.

When shopping for a binocular, put it through a few simple tests in the store. First, try to read some fine print on a distant sign or book to test resolution. This quickly separates the quality models from the cheaper ones. Second, check for flatness of field by viewing something with many lines such as a brick wall or fence. All binoculars will show some distortion or curving in these lines at the edge of the field, but cheap models will have as much as 25 percent of their total field distorted. Third, check the focusing system. Is it smooth, quick, and precise or is it a fight to get something in focus? Fourth, check eye relief if you wear eyeglasses. Reject any model that is uncomfortable or that has too small a field.

Unfortunately, there is no really good test for image brightness under normal store lighting, so you'll have to rely on exit pupil and coating type to evaluate this factor. (The manufacturers' statements are usually quite reliable.)

Don't be in a hurry when you shop. After all, a good binocular is a lifetime investment. Take your time and test as many models as possible before making a decision. A knowledgeable salesperson can be a great help, but you should be prepared to make your own evaluations based on the information in this chapter.

Finally, try to maintain a perspective on quality. If you can afford a premium quality binocular, by all means buy one. You'll be glad in the long run. If you're on a tight budget like most of us, try at least to get into the $100 to $200 range. Remember, the premium quality binocular will get the job done with style, but a basic model will still get the job done.

Chapter 2

OBSERVING DEEP-SKY OBJECTS

An Observing Attitude

It's no secret that many amateur astronomers have a fascination with equipment. There's always a ready market for any product that holds out the promise of better or easier observing. This is a healthy state of affairs as long as it's kept in perspective. In my opinion, though, nothing is sadder than the amateur who mistakenly believes that equipment somehow makes the observer. All too often this sort of thinking produces amateurs who are long on equipment and short on observing skills. (Remember the Parable of Observer A and Observer B.) And make no mistake about it: observing is a skill — one that is developed on the job. Furthermore, it can be learned with any equipment, even with something as small as a 7 x 35 binocular.

For now, though, forget about equipment. Think of observing as a mental activity, something that becomes stronger through repetition and the application of a few mental tools.

Initiative

Observing deep-sky objects is not a spectator sport. A ''sit back and watch'' attitude is the worst possible way to approach this very active and involved process. This doesn't mean you can't relax and enjoy yourself. It simply means that you'll have to show some initiative if you expect to observe many of the deep-sky objects that are visible in a binocular.

Since experience is the key to becoming a good observer, you should be prepared to take advantage of every suitable opportunity to sky-watch. Unless you make observing a priority, it becomes too easy to find something else to do when the weather becomes uncomfortable or you've had a hard day at the office. I know how hard it can be to get away at times, but once you're out under the stars, you'll wonder what took you so long to get there.

As you will likely discover, it really doesn't take that much initiative to become a good observer. Success with a binocular is habit forming. The temptation to pit your observing skills and binocular against a challenging deep-sky object is often too much to resist. Believe me, the hours pass very quickly once you discover how many deep-sky objects are actually visible in a binocular.

There is more to observing, though, than building up an impressive list of objects. It takes a truckload of concentration and plenty of initiative to discern detail in objects that first appear to be nothing more than smudges in your binocular. As an observer, detail should always be your primary goal, no matter what instrument you use.

Initiative is also a useful commodity when you run low on something I call "object appreciation." Even the best observers find that the routine of going out night after night can dull that sense of adventure that is the essence of deep-sky observing. If your sessions begin to go a little stale, you need to rekindle a sense of appreciation and wonder. Use a little initiative to do some extra-credit work on the next objects you plan to observe. There is nothing like knowing an object's true size, distance, and magnitude as you actually look at it to restore your cosmic perspective and cultivate some object appreciation.

Patience and Persistence

Patience and persistence are the real tools of a good observer. They will do more for your skill development than the finest binocular or the darkest observing site. You will need these virtues for all phases of observation, and in ever-increasing quantities as you progress to more difficult objects.

For example, you will have to wait for a night of excellent transparency to detect an object at or near the limiting magnitude of your binocular. Depending on your location, such nights may occur as infrequently as 1 in 10. If you hit a bad stretch of weather or atmospheric haziness, you have little choice but to wait things out. Patience!

Even when conditions are favorable, it usually requires a great deal of persistence to detect a very faint object. My procedure for a difficult object is typical. If my first attempt fails, I recheck its position on my map to make sure I'm looking in the right place. I then sweep the position at least a half-dozen times. If the object still eludes me, I go on to another and come back later. Usually I make two or three of these attempts per session. If the object has not been detected by then, I make a note to try it again on another night. Some objects have taken many nights to find. Some I haven't seen yet. But I will. I'm very patient.

Well, more so on some nights than others. Day-to-day problems can drain my reserve of patience the same as they do for anyone else. (In this respect, at least, deep-sky nuts are like other people.) When this happens I usually try to take a nap before I observe. A nap not only recharges my patience, it also gives my eyes a much-needed rest. I also avoid observing sites that have too many patience-

Checklist for the
Binocular Observer

1) Binocular(s)
2) Tripod
3) Tripod Adapter
4) Sky-Scanner
5) Star Map(s)
6) Notebook
7) Pencils and Sharpener
8) Watch
9) Lens paper and brush

Seasonal:

Insect Repellant
Gloves
Hat
Beverage
Snack

Checklist for the binocular observer.

draining distractions such as passing traffic, hostile crowds, mosquitoes, or excessive noise. I prefer to save my patience for deep-sky objects.

Organization

Nothing separates the expert from the beginner as much as organization. A well-organized observer makes the most of precious observing time and is less likely to overlook a small but potentially important detail.

Organization is largely a matter of preparation — what you do before you reach your observing site. It includes:

1) **A checklist.** It takes so little in the way of equipment to observe with a binocular that a checklist might seem silly. Nevertheless, you're apt to feel worse than silly if you arrive at your site without some essential piece of equipment. It is often some small, easily overlooked item that ultimately does you in. Wait till you forget mosquito repellent on a June night or a pair of gloves in the middle of January.

I keep everything I need to observe in a small overnight bag in a closet. When I'm ready to go out I merely run down my checklist to see if everything is there. Shown opposite is the checklist I use.

As you can see, the list is rather short. However, some items should be explained. *The flashlight* has a red lens since red light is the least disruptive to a dark-adapted eye. *Pencil* rather ballpoint is used because a pencil won't run when wet, won't freeze up when cold, and can be erased easily in case of a mistake. *Gloves* are needed in cold weather and should be used if you get insect repellent on your hands. *Insect repellent can damage the finish on your binoculars and damage the coatings on lenses!* I use the stuff only when absolutely necessary. A *watch* is needed to record the times of celestial events. A *snack* provides needed energy during long sessions. In fact, there is medical evidence that an energy boost can improve your visual acuity.

2) **A good site.** You have to scout for good observing sites. They are rarely found by accident. You need to consider light-pollution intensity and direction, possible obstructions such as trees and buildings, distance from home, access after heavy rain or snow, and possible distractions such as noise, traffic, mosquitoes, etc. Scout by day if possible or on nights of poor seeing. Save good nights for observing.

3) **A nightly plan.** This is perhaps the most important component of organization. It is your nightly strategy. Most beginners make the mistake of observing without a plan or trying to do too much at once. No matter what your experience level, you should decide what to observe before you arrive at your site. Here are some suggestions.

First, concentrate on one small section of the sky each time you observe. This makes it easier to select a favorable site and it also eliminates the need to move constantly and reposition equipment. Moreover, it is easier to learn the sky in small sections than in large chunks.

Second, make a list of objects to observe. Include some bright, easy ones as well as some difficult ones. Don't overdo it. Six new objects per session is plenty if you really stop to look at them. Besides, you can always recheck objects that may have eluded you in a previous session. To be sure, there is never a shortage of these.

Third, always keep a record of your observations. *Don't trust your memory.* There is far too much detail to remember, even in a short session. Describe each object as you observe it. Notes can be brief and to the point, but should always include the following:

a) the date
b) the time
c) an estimate of seeing conditions

d) the location of your site if you have more than one

Record your failures along with your successes. This not only tells you which objects need more work, it also allows you to chart your progress as an observer. Besides, you'll better appreciate your successes when you have a record of what it took to achieve them.

The Observing Site

Your choice of an observing site is more important than your choice of a binocular. Why?

The difference in performance between a 50-mm and an 80-mm binocular amounts to about one magnitude. This may sound like a lot, but it is often less than the night-to-night variation in sky transparency at a good site. It is certainly much less than the three-magnitude gain in performance that can often be realized when moving from a badly light-polluted site to a dark location in the country.

Thus, when you need more performance, it makes more sense to move to a better site rather than trade for a larger instrument. Besides, almost any location can become an observing site when you have a binocular. Remember this the next time you go camping or traveling. In fact, you may even want to plan your next vacation around the dark of the moon.

Rural Sites

Everyone should have the opportunity to observe under a truly dark sky. It's one of the greatest thrills in amateur astronomy. It's also the only way to realize the full potential of your binocular and your observing skills.

Unfortunately, it's becoming harder and harder to find such a site. The light pollution from even a small city spreads out for many miles into the country. Indeed, if you live in a large metropolitan area, you will have to travel a good many miles to find a completely dark observing site.

Fortunately, it isn't necessary to have a completely dark location to do serious work. You can do nearly as well at a light-polluted rural site if the source of the light pollution is in only one direction. In most rural areas near large cities, only one area of the sky is spoiled by ''cityglow.'' The rest of the sky remains dark. Thus it becomes a simple matter of keeping the cityglow behind you and observing in the darkest area of the sky. If you wish to examine southern constellations, select a site to the south of your city. If you wish to observe northern constellations, select a site to the north, and so on. Of course, you'll need several observing sites to cover the entire sky with this strategy, but it is still more practical than driving to a single, more distant location.

My best sites are in the pastures of farmers who have granted me permission to observe, and at public forest and recreation areas that are open at night. I prefer flat or gently rolling grasslands, since these areas have the fewest obstructions and are atmospherically quite stable.

There are several hazards to avoid when selecting a rural site. Number one among these are roads. Roads are a source of light pollution, both from vehicles and traffic signs, and a source of noise pollution as well. They also store up heat during the day and release it slowly at night in the form of image-degrading heat waves. The same can be said for any area devoid of vegetation, such as parking lots or railroad tracks. Be especially cautious of dirt and gravel roads, since the dust kicked up by passing traffic can settle on your equipment and obscure your view.

Observing along a road is also an open invitation for every curiosity seeker or suspicious person to stop and check you out. To be sure, you may make an occasional convert to astronomy, but you may also find that there is precious little time for observing. And speaking from personal experience, I can definitely advise you to inform local residents of your activities when observing along a road if you wish to avoid a question and answer session with a disbelieving local sheriff.

As a last bit of advice, be sure to let someone know when and where you plan to go. The chances of a car breakdown or taking sick while observing are remote, but it is best to be prepared in case of such an emergency. Certainly it is a better plan than trying to hike home or trying to rouse an irate farmer in the middle of the night to make a phone call.

The Suburban Site

Rural sites may be the best, but it is safe to say that more people observe from suburbia. The convenience of observing from a backyard or neighborhood park often outweighs the loss of performance due to light pollution.

And light pollution is a fact of life in the suburbs as well as the city. In general, as I have mentioned before, you may expect to lose from one to three magnitudes of performance when you observe in or near a city. This may seem grim, but there are ways to fight this menace.

One of the best things you can do as a backyard observer is to be on good terms with your neighbors. A polite request that they turn off a yard light or house light is often all it takes to upgrade a backyard site significantly. You may even make some new friends for astronomy if you take the time to show them a few deep-sky objects.

If your neighbors are more the type to throw objects when you make a request, you can take preventive action. Large pieces of cardboard, plywood, or tarpaulin can be erected at strategic locations to block out interfering light. If these ''light barriers'' don't fit into your landscape scheme, you can always rig up a more portable and less permanent model by draping a towel or dark cloth over your head, like an old-fashioned photographer. (You may want to explain this to your neighbors.)

If a backyard site is out of the question, scout around for a vacant lot or a city park. These areas are often the best locations for observing in a city. (Be sure to check local regulations concerning trespassing and legal hours.) Furthermore, they may be within walking or bicycling distance of your home. If you must hop in your car to get to one of these areas, you may as well take the next step and drive on out to a rural site.

Even if you're a housebound or apartmentbound city dweller without a backyard or convenient park nearby, you can still observe some deep-sky objects. Believe it or not, you can do a surprising amount of observing through a closed window. (I learned this several years ago when I was stuck in the house for over a week with the flu.) This technique is impractical for a telescope, which magnifies the defects in window glass, but is quite feasible at the lower power of a binocular. Just remember to darken your room totally and allow your eyes to become dark-adapted. When observing through side windows, it is best to concentrate on objects that are not too high in the sky, since a window causes too much distortion when you look through it at much of an angle. Lucky indeed is the observer who lives in a house or apartment with an overhead window.

Using a Binocular

Focusing. There are several ways to focus a binocular. In my opinion, the following method is the easiest and will produce the least amount of eyestrain.

First, cover the binocular's right side and focus until

stars are sharp in the left eyepiece. This is accomplished by turning the left eyepiece in an individual-focus binocular or the center wheel in a center-focus binocular. Then cover the left side and focus until stars are sharp in the right eyepiece. This is accomplished by turning the right eyepiece in both types of binoculars, though some of the more expensive center-focus models have a separate wheel for focusing the right eyepiece.

Keep both eyes open during focusing. Closing one eye as you focus stretches your eye muscles and produces eyestrain.

Aiming. Most beginners find it difficult to line up their binocular correctly on a star. As a result, they must aim two, three, or four times at a star before it appears in the view. Actually, aiming is a simple process, and can be practiced during daylight hours on any object.

First, just look at the star or object to be viewed. Then bring your binocular up to eye level without moving your head. (Lowering your head is the most common mistake when aiming.) Next, center the star or object to be viewed over the centerpost or hinge of your binocular. This serves as a crude sighting system. Then, without taking your eyes off the star or object, place the eyepieces over your eyes. If done correctly, the star or object should appear in your binocular's field. Practice.

Determining the field of view. One of the great advantages of the binocular is its wide field of view. This makes it possible to locate an object quickly and easily. With experience you'll gain a feel for how much sky your binocular view covers. (It is probably less than you believe at first!) This will enable you to scan large areas of the sky without losing your orientation.

It is wise to measure your binocular's field by checking

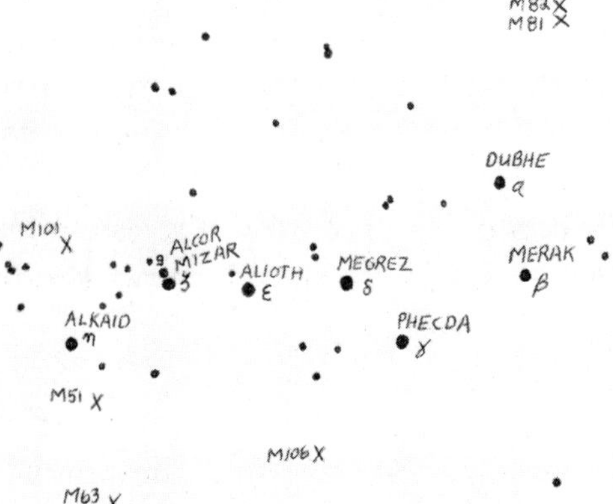

Guide stars in Ursa Major.

it against a star map. Sight on a familiar constellation and see how much area your binocular takes in. The result should be fairly close to the degrees of field labeled on your binocular. (For binoculars labeled in feet at 1,000 yards, 52 feet equals 1°.)

This may not be true if you wear eyeglasses when you observe, especially if your binocular does not have fold-down rubber eyecups (a sign of a short eye-relief model). In some models I have tested, over half of the available field was lost when they were used with eyeglasses. With all the long eye-relief models available today, there is no reason to settle for a reduced field of view if you wear glasses.

Locating Deep-Sky Objects

Before you can detect and observe a deep-sky object, you must first arrive at its position. To get there with a binocular, you need to combine some map-reading skills and a simple technique that uses ''guide'' stars. In essence, this is a simpler and easier version of the star-hopping technique used by telescopic observers.

Begin by locating one or more naked-eye visible stars near the object being sought. The guide star is then centered in the binocular and the object is located by moving the right distance and direction from it — often with the help of lesser stars along the way that are plotted on your map.

The easiest objects to locate are those in the same binocular field as the guide star. Fortunately, a great number of deep-sky objects fall into this category, including M51 with Eta Ursae Majoris (Alkaid) as a guide star in the diagram opposite.

Some objects form convenient triangles with two guide stars. These are only slightly more difficult to locate. An example in the diagram would be M101, which forms an equilateral triangle with guide stars Eta and Zeta Ursae Majoris (Alkaid and Mizar).

The most difficult objects to find are those that require drifting one or more fields away from a guide star or stars. An example, in the diagram, would be M81 and 82, which can be located about 10° north-northwest of Alpha Ursae Majoris (Dubhe). If you follow a line from Phecda through Dubhe and on to M81-82, the job becomes easier. But in such cases you may have to use faint stars plotted on a map to pinpoint the correct location. When you look *exactly* where you should, you can detect objects at least a magnitude fainter than when just sweeping for them.

Convenient guide stars are included in the Location section of the object listings in Chapter 3. These should be considered suggestions, since there is more than one way to locate most objects. Indeed, with a little practice, locating deep-sky objects with guide stars becomes second nature.

General Observing Techniques

Arriving at an object's position is, of course, no guarantee that you'll see it. To detect difficult deep-sky objects consistently, you'll need persistence and sound observing techniques. Furthermore, no amount of expertise will make an object visible if it is beyond the physical limits of your binocular or observing site.

Don't let this discourage you, though. One of the greatest thrills you can experience as a deep-sky observer is finally to detect a difficult object after many attempts. For this reason you should always include a few difficult objects in each observing session. If nothing else, this approach helps to maintain your keen edge as an observer. More importantly, it allows you to accumulate the type of experience that often spells the difference between success or failure on really challenging objects. Here are a few suggestions for spotting an elusive object.

First of all, keep trying! Never draw a conclusion on the basis of one observation. Night-to-night fluctuations in atmospheric conditions and your eye's sensitivity are normal. These variables are often a source of frustration, but they do make observing more interesting.

Second, be sure your eyes are completely dark-adapted. You can begin observing after 15 minutes of darkness, but complete dark adaptation takes half an hour or more. If you get impatient, warm up on some easy objects first, but don't expect your eyes (or your concentration) to be at their best during the first 30 minutes of your observing session.

Third, use averted vision. This is a standard technique for all deep-sky observing and may explain why some amateurs appear cross-eyed. Averted vision is simply looking at an object by gazing a little to its side rather than looking directly at it. This places its image on a more sensitive part of the retina. It's amazing how an object can fade from sight when you look right at it and reappear when you use averted vision.

Fourth, move your binocular slowly back and forth if you fail to spot anything. It's easier to spot a moving object than a stationary one. This technique can be used with a tripod-mounted binocular, but is far more convenient to use with a hand-held one.

Fifth, try looking for a full 6 to 10 seconds without turning away or breaking your gaze when you spot something. Your sensitivity builds continuously as you concentrate on an object.

Sixth, check the published size and magnitude of an object before you attempt to observe it. This gives you a much better idea of what to look for. Should it be a nearly starlike point or a huge, dim cloud? Size and magnitude are supplied for each deep-sky object that is listed in Chapter 3.

Observing Open Star Clusters

The binocular is an excellent instrument for observing open clusters. In fact, its wide field of view makes it a better instrument than the telescope for observing the largest ones. For example, clusters bearing Collinder numbers and other very large open clusters are rarely seen in their entirety when viewed through a telescope. For this reason they are often ignored by telescopic observers. You, however, should not ignore them, because they are interesting objects in a binocular.

As deep-sky objects go, many open clusters are large and bright. For this reason they make excellent targets for the first-time observer. In addition they are available in good numbers throughout the year and can be studied effectively with 7x or 8x binoculars, though 10x or 11x models will resolve more of their stars.

Check the sizes of open clusters carefully before you observe them. They can be as small as 6 or 7 minutes in diameter or cover as much as 5 full degrees of sky. Large, scattered clusters are especially easy to overlook, and are more obvious in a 7x binocular than in a large giant.

You can usually resolve all the stars in these large open clusters. Smaller or denser ones are a different matter. Some of these show individual stars only at the fringes, while others appear as grainy or smooth patches of light with little or no resolution.

There are other things to look for in open clusters. Some reveal subtle differences in the colors of their component stars, while others are uniformly one color from star to star. Perceiving color does take experience, but adds interest to your observations.

You can also see patterns and shapes in most open

clusters. Look especially for lines or rows of stars. You can also observe a wide variety of overall shapes — rectangular, circular, flattened, pointed, etc.

The most difficult open clusters to observe are those that blend into the rich star fields of the Milky Way. To detect these, you'll need to consult a good star map and look for subtle differences in texture that might give a clue to their outlines. Experience pays here.

There are so many open clusters visible in a binocular that you could actually make them a specialty, as some observers have. This is especially true if you concentrate on the lesser-known open clusters such as the Collinders.

Observing Globular Clusters

No commercially available binocular is large enough to resolve these beautiful objects into their component stars very well. Nevertheless, globulars exhibit a surprising degree of individuality. Some appear dense and compact, while others are loose and irregular. Some are large enough to be mistaken for an open cluster, while others appear as little more than a fuzzy star. All in all, they are fine objects for a binocular.

Globulars are a step up in difficulty from the open clusters, since they are, on the average, smaller and dimmer. However, they are still relatively easy and can be found in good numbers.

When observing globulars, check the size and brightness of the central core as well as the degree of resolution at the cluster's fringe. Often you will be able to detect variations in the degree of graininess or fuzziness across the body of the cluster.

Spring and summer offer the best hunting for globulars, but a few are also available in the fall and winter months. Most are in Sagittarius, Scorpius, and Ophiuchus. These are particularly vulnerable to light pollution and poor seeing conditions for northern observers because of their low position in the sky. The 8th-magnitude globular M69 has eluded my 11 x 80's more than once!

Really dim globulars (9th magnitude and fainter) require a good combination of technique and a dark sky. These usually appear as little more than fuzzy stars and are always a challenge to spot. They are an excellent way to sharpen your observing skills before tackling some of the more difficult galaxies.

Observing Galaxies

In this age of large and sophisticated telescopes, the binocular may seem an unlikely tool for spotting galaxies. Nevertheless, a patient observer can see dozens of these distant and faint objects, and even detect some interesting detail in some of the larger ones.

Medium to large galaxies are the easiest targets for binoculars and can be picked up in 7x or 8x instruments by slowly scanning with averted vision. Occasionally you can detect brightness changes across the face of these galaxies and distinguish an elongated shape.

Small galaxies offer less detail at binocular magnifications and are more of a challenge to detect. Generally, galaxies of less than 7′ or 8′ published size are difficult to distinguish from stars in even a giant binocular. Firmly supported 10x to 20x instruments work best for these, but don't sacrifice too much in the way of aperture if you expect to see many galaxies in the 10th- to 11th-magnitude range.

Be prepared to spend some time on the galaxies. They are the ultimate test of your observing skills and your equipment. If they appear unimpressive in a binocular, try to remember what they actually are. Object appreciation! That you can see them at all with something as small as a binocular is amazing.

Observing Planetary Nebulae

With a few notable exceptions, planetaries are less than ideal objects for a binocular. In addition to being dim, most are also rather small — 1′ in size or less. At 7x to 12x, these small planetaries appear more stellar than nebular. M57 (the Ring nebula), for instance, may be spectacular in a large telescope, but in a binocular it appears as little more than a fuzzy star.

Fortunately, a handful of planetaries are both large enough and bright enough to be interesting. In fact, I rate M27 in Vulpecula as one of the best deep-sky objects in a binocular. See if you agree!

Observing Diffuse Nebulae

Although these can be frustratingly faint and elusive, they are still excellent objects for the binocular. Certainly they include some of the most beautiful deep-sky objects.

They also include some of the largest. The North America nebula, for instance, covers some 2 full degrees of sky. The California nebula covers almost 2½° of sky. Think big when you look for these.

Also think faint. Published magnitudes, which treat objects as if all their light were concentrated in a single point, are of little use for objects that have their light spread over so large an area.

To detect faint nebulae, you need to combine a 7-mm exit pupil and the darkest possible sky. The light from these objects is so scattered and ill-defined that the least bit of light pollution or atmospheric haze will wash them out.

There are light-pollution and nebular filters on the market that make it possible to see some very difficult nebulae, such as the California, with the naked eye alone. These filters open up a whole new range of observing possibilities for the binocular observer. If you wish to use them with a binocular, you can either improvise with telescope eyepiece filters by taping them carefully to your binocular or you can use filters made especially for binoculars. These, however, are sometimes hard to find.

Some of the brightest nebulae reveal subtle and beautiful colors on a good night and many show their individual shapes in a binocular. You'll have no trouble, for instance, in tracing the familiar outline of the North America nebula or the Omega nebula, skies permitting. Others, such as the Veil, are wispy and filamentous in nature and require careful scrutiny to untangle their outlines.

Observing Dark Nebulae

These obscuring patches of gas and dust appear as dark blotches or "holes" in the rich star fields of the Milky Way. The dark nebulae, like the largest open clusters, are often ignored by telescopic observers since there is little detail available at telescope magnifications. But they can be observed in good numbers with a binocular. They are especially numerous from Cygnus to the southern reaches of the Milky Way.

Use the same techniques here as with diffuse nebulae. Dark rural skies offer the most contrast with the stellar background, but many dark nebulae can be seen from a good suburban site.

Chapter 3

THE CATALOGUE

Introduction

The following pages list 130 deep-sky objects that are visible in a binocular from north temperate latitudes, including North America and much of Europe. This number is hardly the limit of what can be seen with a binocular, but it does provide a new observer with a good start.

The objects listed here include a cross-section of every type of deep-sky object, ranging in difficulty from very easy to only rarely possible. Most are visible in a small binocular, though some can be seen only with a giant. Many are visible at a good suburban site, but an equal number will probably require a dark rural site. Some are truly a glory to behold, while others are a challenge just to detect.

Instruments

All objects were observed with a minimum of two binoculars, a 7 x 42 and an 11 x 80. These two instruments represent something of a low end and a high end for deep-sky binoculars.

The 7 x 42 was a utility-grade, fully magnesium fluoride coated roof-prism model that retails for about $125. It has

Orion Enhanced 11 x 80 giant binocular. Multi-coatings, BAK-4 prisms, and fine optics make this one of the best deep-sky instruments on the market. Used to collect much of the data in this book.

served me well, and nicely demonstrates what can be done with an ordinary, all-purpose binocular. I feel the 7 x 42 is an excellent compromise between the far more common 7 x 35 and the more light-efficient 7x 50. Observers who use these last two sizes should adjust their expectations accordingly when reading descriptions made with the 7 x 42. For objects not visible in the 7 x 42, the minimum needed size is stated.

The 11 x 80 was a fully multi-coated model employing BAK-4 prisms. Unlike the 7 x 42, it was specifically designed for astronomy, and is optically excellent. Observers who use the smaller giants such as the 8 x 56, the 9 x 63 or the 10 x 70 should adjust their expectations accordingly, but it is my experience that only a little performance is lost when using these smaller instruments.

Observing Site

All observations were made from a variety of rural sites three to five miles outside a city of about 25,000 population. In my area, there is an average of one truly excellent night out of every five observing nights. In terms of darkness, these sites fall somewhere between a light-polluted suburb and a darker, more remote rural site.

The Descriptions

The deep-sky objects are described briefly with the basic information you will need to observe them. There are, however, some important points to remember regarding the data.

Size. The published sizes for deep-sky objects vary somewhat from source to source. More to the point, published sizes are invariably larger than what you actually see in a binocular or amateur telescope, since deep-sky objects are usually measured on highly sensitive photographs. Still, these published sizes are very useful

Swarovski Habicht 8 x 56 Porro-prism binocular. A top-quality instrument that combines tremendous light-efficiency with a unique design.

relative indicators of what you can expect when searching for deep-sky objects.

Magnitude. As with size, published magnitudes are a bit misleading. They usually treat all objects as if they were a single point of light, even though most deep-sky objects have their light spread over a relatively large area. To put it in practical terms, for two objects of the same magnitude, it is usually more difficult to detect the larger of the two. As with published sizes, published magnitudes are useful indicators of what to expect when you observe, but should not be taken too literally.

Location. It is especially helpful to know the width of your binocular's field when checking this category since most directions are given in degrees from one or more guide stars. Remember that north is always the direction to Polaris, and east is 90° counterclockwise from there. That is, if your binocular field of view were a clock face with 12 o'clock toward Polaris, east would be 9 o'clock, south 6 o'clock, and west 3 o'clock. The guide star listed was usually the most convenient for me, but you may find that other stars work better for you. Experiment.

Comments. With some reservations, I decided to keep my comments brief and to the point. In my experience, long, detailed, and elaborate descriptions often confuse and intimidate new observers. Furthermore, you should be more concerned with what you see when *you* observe rather than trying to ''resee'' what I did. Besides, much of the fun in observing is describing and seeing objects in your own way.

Dark skies to all!

DEEP-SKY OBJECTS FOR BINOCULARS

ANDROMEDA

NGC 752
Type: Large, irregular open cluster of about 60 or 70 stars
Size: 45′
Mag.: 6
Location: 5° south of Gamma (γ) Andromedae
Comments: Be sure to check NGC 752 when you're in the area to observe more famous objects such as M31 and M33. A 7 x 42 reveals a dusting of faint stars against a milky background. An 11 x 80 resolves dozens of stars in this beautiful object.

M31, ANDROMEDA GALAXY
Type: Large spiral galaxy, visible to the naked eye, and a member of our Local Group; our next-door neighbor at 2.2 million light-years
Size: 160′ × 40′
Mag.: 4.8
Location: 1° west of Nu (ν) Andromedae
Comments: One of nature's grandest sights in a binocular. Appears as a huge ball of milky light in a 7 x 42. Fills the better part of a field in a giant binocular. An 11 x 80 reveals a bright, compact hub surrounded by a huge, ill-defined disk.

AQUARIUS

M2
Type: Rich globular cluster
Size: 12.9′
Mag.: 6.5
Location: 5° north of Beta (β) Aquarii, almost on the line to Epsilon (ϵ) Pegasi
Comments: One of the best globulars in the sky. Fine object in any binocular. Compare with M5 and M13.

NGC 7293, HELICAL NEBULA
Type: Large planetary nebula, the largest in the sky
Size: 13′
Mag.: 6.5
Location: About 8° southwest of Delta (δ) Aquarii, 1° west of Upsilon (υ) Aquarii
Comments: One of a handful of planetaries large enough to appear as more than a star in a binocular. Don't let the published magnitude of 6.5 fool you. The light from the Helical is spread over so large an area that it tends to blend in with the sky background. On an excellent night, the Helical tends to fade in and out of a 7 x 42. An 11 x 80 consistently picks up a patch of haze 10′ across. Often missed by telescopic observers.

AQUILA

B142, B143
Type: Dark nebulae
Size: 80′ × 50′
Location: 2° west of Gamma (γ) Aquilae
Comments: This interesting object is a double cloud of obscuring gas and dust that appears as a large, slanted letter *E* on an excellent night in the rich star field west of Gamma (γ) Aquilae. The top two prongs are B143 and the bottom prong is B142. The B in B142 and B143 is for Barnard, the astronomer who cataloged these nebulae. Guess what his first two initials are?

Open star clusters in Auriga are fine binocular targets. Photograph by Dennis di Cicco with a 110-mm f/2.8 lens.

AURIGA

M38
Type: Large open cluster of about 100 stars
Size: 21′
Mag.: 6.4
Location: Just north of a line between Theta (θ) and Iota (ι) Aurigae, midway between them; this is the northernmost of the three famous clusters in Auriga (M38, M36, M37)
Comments: This is the easiest of the three famous clusters to resolve with a binocular. Some observers report a cross or U shape for this fine object.

M36
Type: Small, bright open cluster of about 60 stars
Size: 12′
Location: Just south of a line between Theta (θ) and Iota (ι) Aurigae, slightly closer to Theta; M38 is only 2° to the northwest
Comments: This cluster is smaller and slightly brighter than M38. It provides an interesting contrast to the much larger clusters above and below it.

M37
Type: Large and very rich open cluster of more than 150 stars
Size: 24′
Mag.: 5.6
Location: About 3½° southeast of M36
Comments: A superb object in a binocular. A 7 x 42 reveals a hazy patch of light with a scattering of stars. An 11 x 80 resolves many stars across the face of the cluster. A classic.

NGC 1664
Type: Open cluster of about 40 stars
Size: 18′
Mag.: 7.6
Location: About 2° west of Epsilon (ϵ) Aurigae
Comments: Small and faint in a 7 x 42, but resolvable. Better object in an 11 x 80.

NGC 1893
Type: Irregular open cluster of about 20 stars
Size: 11′
Mag.: 7.5
Location: About 2½° west-southwest of M36
Comments: A small, inconspicuous cluster in the same field as M36, near 16, 17, 18, and 19 Aurigae.

NGC 2281
Type: Open cluster of 30 stars
Size: 15′
Mag.: 7
Location: About 17° east-southeast of Capella; 1° south-southwest of Psi^7 (ψ^7) Aurigae
Comments: A small, loose cluster, but fairly bright. Easy object for a 7 x 42.

BOOTES

NGC 5466
Type: Globular cluster
Size: 5′
Mag.: 9
Location: About 10° north and a little west of Arcturus; about 1½° northeast of 11 Bootis
Comments: A very difficult object in a 7 x 42. Look for a tiny ball of light on a very dark night with an 11 x 80.

CAMELOPARDALIS

NGC 2403
Type: Spiral galaxy
Size: 16′ × 10′
Mag.: 8.4
Location: About 7½° northwest of Omicron (*o*) Ursae Majoris, almost in a line with Theta (θ) Ursae Majoris; 1° west of 51 Camelopardalis
Comments: Moderately difficult object of very low surface brightness. Can be picked up in a 7 x 42 under excellent conditions, but usually requires something larger. If you find this object unimpressive, remember it is 8 million light-years distant!

NGC 1502
Type: Open cluster of about 45 stars
Size: 8′
Mag.: 5.7
Location: Forms the apex of an equilateral triangle with Alpha (α) and Beta (β) Camelopardalis, to their west
Comments: Small but rich open cluster bright enough to be seen in any binocular. Fine object in an 11 x 80.

CANCER

M44, BEEHIVE
Type: Very large open cluster of 50 stars
Size: 100′
Mag.: 3.1
Location: Easily found $^3/_5$ of the way from Regulus to Pollux
Comments: A binocular classic. This cluster is too large to fit in the field of the average telescope. A 7 x 42 resolves dozens of stars of varying colors.

M67
Type: Compact open cluster of 200 stars
Size: 30′
Mag.: 6.9
Location: About 8° south-southwest of the Beehive and less than 2° west of Alpha (α) Cancri
Comments: Quite a contrast to the Beehive, with eight times as many stars compacted into a third the size. A 7 x 42 reveals a rich misty patch of light sprinkled with tiny specks of stars. An 11 x 80 resolves many stars across the face of the cluster. Superb object.

The Beehive cluster in Cancer is a splendid binocular object. Photograph by Dennis di Cicco.

CANES VENATICI

M106
Type: Large spiral galaxy
Size: 18′ × 18′
Mag.: 8.3
Location: A little less than halfway from Beta (β) Canum Venaticorum to Gamma (γ) Ursae Majoris
Comments: Even a 7 x 42 shows this elongated disk. An 11 x 80 reveals a bright central hub surrounded by a nebulous disk. Fine galaxy for a binocular.

M94
Type: Large spiral galaxy
Size: 11′×9′
Mag.: 8.2
Location: Forms the apex of a very shallow isoceles triangle, only 1½° high, with Alpha (α) and Beta (β) Canum Venaticorum; the galaxy is to their northeast
Comments: Somewhat more difficult than its magnitude suggests, but still possible with a 7 x 42 on the best of nights. A bright central hub surrounded by a very faint disk in an 11 x 80.

M63
Type: Spiral galaxy
Size: 12′×7.6′
Mag.: 8.6
Location: About a third of the way from Alpha (α) Canum Venaticorum to Alkaid, the end star in the Big Dipper's handle; about 1½° north of four faint stars — 18, 19, 20 and 23 Canum Venaticorum
Comments: An elongated smudge visible in a 7 x 50 on a good night. You may pick up an 8th-magnitude star near the galaxy's western edge.

M51, WHIRLPOOL GALAXY
Type: Large spiral galaxy
Size: 11′×7.8′
Mag.: 8.4
Location: About a quarter of the way from Alkaid, the end star in the Big Dipper's handle, to Alpha (α) Canum Venaticorum
Comments: A large nebulous disk even in a 7 x 42. Perhaps the best galaxy in Canes Venatici for a binocular.

M3
Type: Globular cluster
Size: 10′
Mag.: 6

Location: 12° northwest of Arcturus, on a line to Alpha (α) Canum Venaticorum
Comments: A fine object in any binocular. This globular and M53 to the south in Coma Berenices provide a nice change of scenery in an area otherwise dominated by galaxies.

CANIS MAJOR

M41
Type: Large open cluster of 80 stars
Size: 38′
Mag.: 4.5
Location: 4° south of Sirius; forms a triangle with Sirius and Beta (β)
Comments: A large glowing patch of sky in a 7 x 42. An 11 x 80 resolves many of the cluster's stars and reveals some interesting patterns. (I often take a break from studying or writing by propping my binocular on my bedroom window and observing M41. It sure beats watching TV.)

COLLINDER 121
Type: Large scattered open cluster of about 20 stars
Size: 50′
Mag.: 2.5
Location: Known as the Omicron (ο) Canis Majoris cluster, since it is centered on that star
Comments: Collinder clusters are typically large and scattered — ideal objects for a binocular. Be sure to check the size. This one is a nice object in a 7 x 42.

COLLINDER 132
Type: Large scattered open cluster of about 25 stars
Size: 95′
Mag.: 3.5
Location: About 3° southwest of Eta (η) Canis Majoris
Comments: Check the size of this cluster. Conspicuous in a 7 x 50.

COLLINDER 140
Type: Very large scattered open cluster of about 30 stars
Size: 42′
Mag.: 3.5
Location: About 3° south of Eta (η) Canis Majoris
Comments: A wide, very loose cluster best viewed in a 7x or 8x binocular.

NGC 2362
Type: Small open cluster of about 40 stars
Size: 6′
Mag.: 6
Location: Surrounds Tau (τ) Canis Majoris
Comments: Provides an interesting contrast to the Collinder objects. Appears as a nebulosity around Tau in a 7 x 42. An 11 x 80 shows it to be a small compact cluster.

CAPRICORNUS

M30
Type: Globular cluster
Size: 11′
Mag.: 8
Location: 3° east and a little south of Zeta (ζ) Capricorni; $^2/_5$° west of 41 Capricorni
Comments: Can be an elusive object because of its low position in the sky. An unresolvable ball of fuzz even in an 11 x 80.

CASSIOPEIA

M52
Type: Open cluster of over 100 stars
Size: 13′
Mag.: 7
Location: Found on a line extended from Alpha (α) through Beta (β) Cassiopeiae at a distance of slightly more than the separation of the two stars
Comments: A small but bright open cluster. Nice object in any binocular.

NGC 663
Type: Open cluster of 80 stars
Size: 11′
Mag.: 7
Location: Just southeast of the point halfway between Delta (δ) and Epsilon (ϵ) Cassiopeiae
Comments: Nice cluster in any binocular. A 7 x 42 reveals a speckled surface on a misty background.

NGC 7789
Type: Very rich open cluster of 300 stars
Size: 16′
Mag.: 6.7
Location: 3° southwest of Beta (β) Cassiopeiae
Comments: Appears as a large, milky white patch in all binoculars. A splendid object even if not resolvable.

M103
Type: Small open cluster of 25 stars
Size: 6′
Mag.: 7.4
Location: Only 1° northeast of Delta (δ) Cassiopeiae
Comments: Very small and inconspicuous cluster. Easily overlooked, but visible in a 7 x 42.

NGC 457
Type: Open cluster of 80 stars
Size: 13′
Mag.: 6.4
Location: Extends northwest from Phi (ϕ) Cassiopeiae; 2° south-southwest of Delta (δ) Cassiopeiae
Comments: A small, nebulous appearing cluster in a 7 x 42. Phi Cassiopeiae appears inside this cluster but is not a cluster member. An 11 x 80 resolves many of the cluster's members.

MELOTTE 15
Type: Open cluster of 40 stars
Size: 22′
Mag.: 6.5
Location: 5° southeast of Epsilon (ϵ) Cassiopeiae; 5° north-northwest of the Perseus Double Cluster
Comments: Small cluster that resolves nicely in an 11 x 80.

CETUS

NGC 246
Type: Planetary nebula
Size: 4′
Mag.: 8.5
Location: 6° north of Beta (β) Ceti; forms an equilateral triangle with Phi¹ (ϕ^1) and Phi² (ϕ^2) Ceti, south of them
Comments: More difficult than its 8.5 magnitude suggests due to its low position in the sky. Just large enough to appear as a small globe in an 11 x 80.

COMA BERENICES

MELOTTE 111, COMA STAR CLUSTER
Type: Huge open cluster of 47 stars
Size: 6° (not minutes)
Mag.: 2.9
Location: Look for this scattered north-south cluster between Arcturus and the tail of Leo
Comments: Because of its size, this cluster can easily be found with the naked eye. You'll need a binocular with a field of at least 7° to get the best view of this immense cluster.

M53
Type: Large globular cluster
Size: 14′
Mag.: 8.7
Location: 1° northeast of the double star Alpha (α) Comae Berenices
Comments: Faint object in a 7 x 42, but no problem in an 11 x 80. Fine globular.

M64
Type: Spiral galaxy
Size: 7.5′ × 3.5′
Mag.: 8.6
Location: 5° northwest of Alpha (α) Comae Berenices, 1° east of the star 35 Comae
Comments: A small but bright galaxy. Appears as a small fuzzy star in a 7 x 42. An 11 x 80 reveals more of the galaxy's disk shape.

CYGNUS

M29
Type: Small open cluster of 20 stars
Size: 7′
Mag.: 7
Location: 1½° south of Gamma (γ) Cygni, the center star in the Northern Cross
Comments: A speckled patch of light in a 7 x 42. An 11 x 80 resolves the cluster.

M39
Type: Large open cluster of 25 stars
Size: 30′
Mag.: 5
Location: 9° east-northeast of Deneb
Comments: Just right for a 7 x 42 or larger instrument. Loose cluster easily resolved.

NGC 6910
Type: Small open cluster of 40 stars
Size: 8′
Mag.: 6.5
Location: 3/5° north-northeast of Gamma (γ) Cygni
Comments: Not an obvious cluster. Appears as a patch of nebulous light near Gamma in a 7 x 42. This small Y-shaped cluster is associated with an emission complex known as IC 1318.

NGC 7000, NORTH AMERICA NEBULA
Type: Large emission nebula
Size: 100′ × 120′
Location: 3° east of Deneb
Comments: Bright enough to be seen in any binocular. The North America shape is obvious, especially the Gulf of Mexico and the Pacific coast. You may also pick up a portion of the Pelican nebula to its west in the same field.

IC 5067, PELICAN NEBULA
Type: Large emission nebula
Size: 85′ × 75′
Location: Appears as an island out in the Atlantic near the North America nebula
Comments: Much dimmer than the North America. A 7 x 50 or other binocular with a 7-mm exit pupil will reveal two lobes in this nebula on a very dark night. The lobes can be found along both sides of a line connecting the stars 56 and 57 Cygni.

NGC 6992-5, NGC 6960, VEIL NEBULA
Type: Emission nebula
Size: 78′×8′ (NGC 6992-5); 70′×6′ (NGC 6960)
Location: NGC 6992-5, the eastern section of the nebula, can be found about 3° south and a little west of Epsilon (ϵ) Cygni; NGC 6960, the western section, can be found 2½° to the west, bisected by the star 52 Cygni
Comments: These two long, wispy filaments of light are visible in a 7 x 50 or a giant binocular on the very best nights. The eastern section is brighter, but the western section is easier to find because it is centered on 52 Cygni. Use nebular filters if you have them.

DRACO

NGC 4236
Type: Barred spiral galaxy
Size: 18.6′ × 6.9′
Mag.: 9.7
Location: About 11° north of the bowl of the Big Dipper, about 1½° west-southwest of Kappa (κ) Draconis
Comments: Possible in a 7 x 50 on the best of nights, but usually a giant binocular is required to find this 10th-magnitude smudge.

GEMINI

M35
Type: Large, rich open cluster of 120 stars
Size: 30′
Mag.: 5.5
Location: About 2° northwest of Eta (η) Geminorum
Comments: Fine cluster for any binocular. A 7 x 42 shows a large patch of light dotted with stars. An 11 x 80 reveals a magnificently rich and well-resolved cluster.

NGC 2129
Type: Small open cluster of 50 stars
Size: 5′
Mag.: 6.7
Location: Can be found in the same field as M35; a much smaller glow to the south and west, ⅔° west of 1 Geminorum
Comments: A very small, faint glow in a 7 x 42. An interesting contrast to M35.

HERCULES

M13
Type: Large, rich globular cluster
Size: 16.6′
Mag.: 5.9
Location: A third of the way along a line from Eta (η) to Zeta (ζ) Herculis
Comments: The most famous globular in the sky and a fine object in any binocular. Unfortunately, not even an 11 x 80 does much to resolve the core of the cluster. It can, however, resolve some stars at the fringes on a good night. Enjoy.

M92
Type: Large globular cluster
Size: 11′
Mag.: 6.5
Location: $^2/_5$ of the way from Iota (ι) to Pi (π) Herculis; forms an equilateral triangle with Theta (θ) and Rho (ϱ) Herculis
Comments: Often overlooked in favor of M13, but a fine globular even in a 7 x 42.

HYDRA

M48
Type: Large open cluster of 80 stars
Size: 30′
Mag.: 5.9
Location: 3° south and a little east of Zeta (ζ) Monocerotis (not 4° to the north as Messier reported)
Comments: A 7 x 42 reveals a triangular shape and resolves many stars. Nice cluster.

M68
Type: Globular cluster
Size: 8′
Mag.: 8
Location: 4° south and a little east of Beta (β) Corvi
Comments: Tiny ball of light in a 7 x 42. Usually requires careful scanning on a dark night because of its low position in the sky.

M83
Type: Spiral galaxy
Size: 8′ × 10′
Mag.: 9
Location: Slightly more than halfway from Gamma (γ) Hydrae to Theta (θ) Centauri
Comments: Difficult object because of its low position and because its light is spread over a large area. Requires the best sky transparency. Possible with a 7 x 50 under exceptional conditions. Usually requires a giant.

LACERTA

NGC 7243
Type: Open cluster of about 40 stars
Size: 20′
Mag.: 7
Location: 2½° west and a little south of Alpha (α) Lacertae
Comments: A 7 x 42 reveals a milky haze that curves slightly northward. An 11 x 80 resolves some of the cluster's stars, especially at the fringes.

NGC 7209
Type: Open cluster of about 50 stars
Size: 26′
Mag.: 7.8
Location: 6° southwest of Alpha (α) Lacertae
Comments: A dim patch of light in a 7 x 42. An 11 x 80 resolves some stars across the face of the cluster.

LEO

M65, M66
Type: Spiral galaxies
Size: 7.8′ × 1.6′ (M65); 8′ × 2.5′ (M66)
Mag.: 10.2 (M65); 9.7 (M66)
Location: Halfway between Theta (θ) and Iota (ι) Leonis
Comments: These two galaxies are separated by only 21′. On a very good night they are visible in a 7 x 50 if you use averted vision. An 11 x 80 does a much better job of picking them up and separating them into two distinct objects. On an exceptional night you may be able to spot a 10th-magnitude edge-on galaxy just north of M65. This thin sliver of light is NGC 3628. For an added challenge, try M95 and M96 located 8° to the west.

NGC 2903
Type: Spiral galaxy
Size: 11′ × 4.7′
Mag.: 9.5
Location: 1½° south of Lambda (λ) Leonis
Comments: Elusive object; can be picked up in a 7 x 50 as a large, elongated smudge on a good night. An 11 x 80 is just right for this object.

LEPUS

M79
Type: Globular cluster
Size: 8′
Mag.: 8.4
Location: Forms a nearly equilateral triangle with Beta (β) and Epsilon (ϵ) Leporis, to their south
Comments: One of the few globulars visible during the winter months. A difficult object for a 7 x 42 because of its low position in the sky. Requires careful scanning even with an 11 x 80.

LYRA

M56
Type: Globular cluster
Size: 7′
Mag.: 8.3
Location: Almost halfway from Beta (β) Cygni to Gamma (γ) Lyrae
Comments: A small, fuzzy dot in a 7 x 42, but an obvious cluster in an 11 x 80.

MONOCEROS

NGC 2244
Type: Bright open cluster of 100 stars
Size: 24′
Mag.: 4.8
Location: About 8° east and a little south of Betelgeuse, 2° east of the star Epsilon (ϵ) Monocerotis
Comments: A milky-appearing cluster in a 7 x 42, embedded with stars in an 11 x 80. This cluster is actually surrounded by the Rosette nebula, which is itself a very large, but difficult to detect, object in a binocular.

NGC 2353
Type: Large, scattered open cluster of about 30 stars
Size: 20′
Mag.: 7
Location: 5° east and a little north of Theta (θ) Canis Majoris
Comments: Easy to overlook. A 7 x 42 reveals a loose, widely scattered cluster. Easily resolved in an 11 x 80.

M50
Type: Bright open cluster of 80 stars
Size: 15′
Mag.: 6
Location: About 9° north-northwest of Sirius on a line with Procyon
Comments: Nice cluster. Appears as a bright smudge in a 7 x 42, but resolves nicely in an 11 x 80.

OPHIUCHUS

M62
Type: Globular cluster
Size: 14′
Mag.: 6.5
Location: 5° north-northeast of Epsilon (ϵ) Scorpii
Comments: A superb globular that suffers somewhat from its low position in the sky. A 7 x 42 shows a whitish ball of light, while an 11 x 80 resolves some of the stars at its fringes.

M9
Type: Globular cluster
Size: 9′
Mag.: 7.8
Location: A little over 3° southeast of Eta (η) Ophiuchi
Comments: Appears as a small ball of fuzz in a 7 x 42. A giant reveals the true nature of the object.

M10
Type: Large globular cluster
Size: 15′
Mag.: 6.6
Location: 1° west of 30 Ophiuchi; forms an equilateral triangle with Zeta (ζ) and Epsilon (ϵ) to the east and southeast
Comments: A glowing ball of light in a 7 x 42. Fine object in an 11 x 80.

M12
Type: Large globular cluster
Size: 14.5′
Mag.: 6.9
Location: Only 3° northwest of M10; drift 8° east-northeastward from Delta (δ) and Epsilon (ϵ) Ophiuchi
Comments: A twin to M10. Fine object. Picked your favorite yet?

M14
Type: Globular cluster
Size: 12′
Mag.: 7.5
Location: About 10° east of M12, or 8° south and a little west of Beta (β) Ophiuchi
Comments: Not as impressive as M10 and M12, but still an interesting object in a 7 x 42. Better yet in an 11 x 80.

M107
Type: Globular cluster
Size: 10′
Mag.: 8.2
Location: About 3° south and a little west of Zeta (ζ) Ophiuchi
Comments: The dimmest of the Ophiuchus binocular globulars, but still possible with a 7 x 42 on a very good night. Much easier with an 11 x 80.

IC 4665
Type: Large open cluster of 20 stars
Size: 55′
Mag.: 6
Location: 1½° northeast of Beta (β) Ophiuchi
Comments: One of those open clusters too large for a telescope. A 7 x 42 resolves the sparse scattering of stars.

M19
Type: Globular cluster
Size: 13.5′
Mag.: 6.8
Location: 7½° east of Antares
Comments: Not as condensed as nearby M62, but still a fine sight in a binocular. You'll need a 7 x 50 or larger binocular to find this dim, low-to-the-horizon globular.

ORION

M42, M43; ORION NEBULA
Type: Emission nebula
Size: 85′ × 60′ (M42); 5′ × 10′ (M43)
Location: Surrounds the muiltiple star Theta (θ) Orionis
Comments: Visible without a binocular on a good night, but you'll need at least a 7 x 50 to let the nebula show its stuff. An 11 x 80 will separate the northern branch of the nebula (M43) from the main body (M42).

M78
Type: Reflection nebula
Size: 8′ × 6′
Location: About a fourth of the way from Zeta (ζ) Orionis to Betelgeuse
Comments: Just beyond the reach of a 7 x 42. A small, indefinite patch of nebulosity in an 11 x 80. Unusually small nebula.

PEGASUS

M15
Type: Globular cluster
Size: 12′
Mag.: 6.5
Location: 4° northwest of Epsilon (ϵ) Pegasi
Comments: A 7 x 42 picks up the cluster as a large, glowing ball of fuzz. An 11 x 80 reveals a ball of bright light. Excellent object in an otherwise barren area.

PERSEUS

NGC 869, NGC 884; THE DOUBLE CLUSTER
Type: Large open clusters of 300 stars or more
Size: 35′ for both clusters
Mag.: 4.4 (NGC 869); 4.7 (NGC 884)
Location: Follow a line from Gamma (γ) through Delta (δ) Cassiopeiae for a distance of about twice the separation between the two stars
Comments: Visible with the naked eye on a good night and a grand sight in a binocular on any night. A 7 x 42 reveals dozens of stars in each cluster. An 11 x 80 picks up a few red stars among the mostly blue and white ones. Spend some time on this cluster.

NGC 1528
Type: Open cluster of 80 stars
Size: 25′
Mag.: 6
Location: North of the line between Capella and Alpha (α) Persei, about halfway between them; 1½° northeast of Lambda (λ) Persei
Comments: A faint, curving row of stars against a milky background in a 7 x 42. Better resolved in an 11 x 80.

NGC 1245
Type: Open cluster of about 40 stars
Size: 20′
Mag.: 6.9
Location: About a third of the way from Alpha (α) Persei to Algol
Comments: A slightly elongated group that is much resolved even in a 7 x 42. Excellent object in an 11 x 80.

NGC 1342
Type: Scattered open cluster of about 40 stars
Size: 15′
Mag.: 7
Location: About halfway along a line between Algol and Zeta (ζ) Persei
Comments: Faint haze with a sprinkling of stars in a 7 x 42. An 11 x 80 reveals a more obvious scattering of stars.

NGC 957
Type: Open cluster of 40 stars
Size: 10′
Mag.: 7.2
Location: Almost 2° east of the Double Cluster; 3° northwest of Eta (η) Persei
Comments: Considerably dimmer than the Double Cluster, but well within the grasp of a 7 x 42 as a faint misty patch. A giant binocular will resolve much of the cluster at its fringe.

M34
Type: Open cluster of 80 stars
Size: 20′
Mag.: 5.5
Location: About 2/5 of the way from Algol to Gamma (γ) Andromedae and a little north
Comments: Excellent open cluster for binocular. A 7 x 42 reveals a hazy central core sprinkled with stars. An 11 x 80 outlines the smaller central core and picks up many stragglers around the fringes.

NGC 1499, CALIFORNIA NEBULA
Type: Large emission nebula
Size: 145′ × 40′
Location: About 1/5 of the way from Xi (ξ) to Epsilon (ϵ) Persei
Comments: A challenge for any binocular. Use nebular filters if you have them. I pick it up as a ''positive'' about one night in five with my 11 x 80, and as a ''maybe'' about one night in three. This is the sort of object that makes observing so exciting.

MELOTTE 20, ALPHA PERSEI GROUP
Type: Very large stellar association of about 70 stars
Size: 3° (not minutes)
Mag.: 1.2
Location: Includes the star Alpha (α) Persei
Comments: Too large to be seen at its best in a telescope, but just right for a binocular. Beautiful and impressive sight in any binocular, but fits perfectly in an 11 x 80.

PISCES

M74
Type: Large spiral galaxy
Size: 9.5′ × 10.2′
Mag.: 9.2
Location: 1½° east-northeast of Eta (η) Piscium
Comments: Fades in and out with a 7 x 42 on a good night. An 11 x 80 easily captures the galaxy as a small round smudge.

PUPPIS

M46
Type: Rich open cluster of 150 stars
Size: 25′
Mag.: 8
Location: In the same field as M47 and NGC 2423. Just south of a line extended from Beta (β) Canis Majoris through Sirius for about 2½ times the separation of the two stars
Comments: This is the richest of the three clusters. Appears as a soft glow in a 7 x 42, with some resolution in an 11 x 80.

M47
Type: Open cluster of 25 stars
Size: 20′
Mag.: 5.5
Location: 1½° west of M46 in the same field
Comments: Brighter and better resolved than M46 in a 7 x 42. Fine object.

NGC 2423
Type: Open cluster of about 60 stars
Size: 20′
Mag.: 9.5
Location: Less than 1° north of M47
Comments: Usually beyond the grasp of a 7 x 42. Appears as a very dim northern extension of M47 in an 11 x 80. An interesting effect.

NGC 2477
Type: Rich open cluster of 50 stars
Size: 18′
Mag.: 7
Location: About 2½° northwest of Zeta (ζ) Puppis
Comments: Very low in the sky for northern observers, but well worth observing when conditions permit. A rich, milky patch of light in a 7 x 42 on a good night; some resolution in an 11 x 80.

M93
Type: Open cluster of about 50 stars
Size: 18′
Mag.: 7
Location: 9° south of M46; 7° northeast of Eta (η) Canis Majoris
Comments: A small patch of misty light in a 7 x 42. An 11 x 80 reveals a milky background dotted with tiny stars.

SAGITTARIUS

M54
Type: Globular cluster
Size: 6′
Mag.: 9
Location: Nearly 2° west-southwest of Zeta (ζ) Sagittarii, almost on a line with Epsilon (ϵ)
Comments: Faint globular found low in the sky. Difficult object. Nice challenge for an 11 x 80.

M23
Type: Rich open cluster of 120 stars
Size: 25′
Mag.: 7
Location: About 10° south of Nu (ν) Ophiuchi, west of the line to Gamma (γ) Sagittarii
Comments: Beautiful cluster in any binocular. Bright patch of light dotted with many stars in a 7 x 42. Mostly resolved with an 11 x 80. Reminds me of M41 in Canis Major.

M20, TRIFID NEBULA
Type: Emission nebula
Size: 29′ x 27′
Location: 7½° north of Gamma (γ) Sagittarii on a line with Nu (ν) Ophiuchi; M8 is 1½° south and M21 is less than 1° northwest
Comments: Fine object in a 7 x 50 under a dark sky. Can you see the dark intersecting lanes that divide the nebula?

M8, LAGOON NEBULA
Type: Emission nebula
Size: 80′ x 35′
Location: 6° north of Gamma (γ) Sagittarii on a line with Nu (ν) Ophiuchi; just to the south of M20
Comments: Superb sight in a 7 x 50 or larger instrument. The bright open cluster NGC 6530 is surrounded by the nebula.

M21
Type: Compact open cluster of 50 stars
Size: 12′
Mag.: 7
Location: ⅔° northeast of M20
Comments: Small patch of light studded with many stars in a 7 x 42. Superb sight in an 11 x 80.

M24, SMALL SAGITTARIUS STAR CLOUD
Type: Large detached portion of the Milky Way, separated by dark nebulae
Size: 120′ × 40′
Mag.: 4.5
Location: 7½° north of Lambda (λ) Sagittarii
Comments: An incredibly thick piece of the Milky Way. Too large for a telescope but just right for a binocular. A grainy cloud of stars in a 7 x 42.

B92
Type: Dark nebula
Size: 15′
Location: Northwestern corner of M24 (Small Sagittarius Star Cloud)
Comments: Appears as a dark hole in the bright star cloud. Most distinct on nights of excellent transparency.

M18
Type: Sparse, small open cluster of 12 stars
Size: 7′
Mag.: 8
Location: 8° north and a little west of Lambda (λ) Sagittarii; in the same field as M17 and M24
Comments: Too loose to show much in a telescope, but just right for a binocular. Faint, but resolvable in a 7 x 42.

M17, OMEGA NEBULA
Type: Emission nebula
Size: 46′ × 37′
Mag.: 6.9
Location: 9° north of Lambda (λ) Sagittarii, 1° north-northeast of M18
Comments: One of the finest emission nebulae in the sky. The "omega" shape is visible even in a 7 x 42. Some observers see a number 2 shape. Either way a fine object.

M22
Type: Globular cluster
Size: 18′
Mag.: 6
Location: 2½° northeast of Lambda (λ) Sagittarii
Comments: Large, milky ball of light in any binocular. Impressive object.

M28
Type: Globular cluster
Size: 6′
Mag.: 8
Location: 1° northwest of Lambda (λ) Sagittarii
Comments: Much dimmer than nearby M22, but no problem for a 7 x 50 on a good night, other than its small size.

M25
Type: Large open cluster of 50 stars
Size: 35′
Mag.: 6
Location: 6° north and a little east of Lambda (λ) Sagittarii
Comments: Nice open cluster in any binocular. Many stars resolved even with a 7 x 42.

M55
Type: Large globular
Size: 15′
Mag.: 7
Location: On a line extended from Sigma (σ) through Tau (τ) Sagittarii, about 2½ times the separation between the two stars
Comments: A fine globular located too far south to really show its stuff. On a good night you may resolve some stars with an 11 x 80.

M70
Type: Globular cluster
Size: 4′
Mag.: 8
Location: Halfway between Zeta (ζ) and Epsilon (ε) Sagittarii
Comments: Slightly brighter but even smaller than M28. Visible in a 7 x 50 under the best conditions. Easier in a 10x or 11x.

M69
Type: Globular cluster
Size: 4′
Mag.: 7.5
Location: A little more than ⅔ of the way from Zeta (ζ) to Epsilon (ε) Sagittarii and somewhat north
Comments: Tougher than its magnitude suggests. Usually difficult for a 7 x 50 because of its size and low position. Easier in a 10x or 11x.

SCORPIUS

M7
Type: Magnificent open cluster of 130 stars
Size: 50′
Mag.: 3.3
Location: 4½° east-northeast of Lambda (λ) Scorpii in the Scorpion's tail
Comments: Perhaps the best open cluster in the sky in a binocular. Almost seems alive in a 7 x 50 or larger instrument. Any binocular resolves dozens of stars. A giant can resolve stars to the very core of the cluster.

M6
Type: Bright open cluster of 132 stars
Size: 26′
Mag.: 4.6
Location: About 4° northwest of M7; 5° west and a little south of Gamma (γ) Sagittarii
Comments: Hard to believe that two such incredible clusters can be found only 4° apart. Many observers report a butterfly shape.

M4
Type: Large globular cluster
Size: 22′
Mag.: 6.4
Location: Only 1⅕° west of Antares
Comments: Visible in a 7 x 42 as a large, fuzzy ball of light. An 11 x 80 can resolve stars at the fringes of this cluster.

M80
Type: Small globular cluster
Size: 7′
Mag.: 8.4
Location: Halfway between Antares and Beta (β) Scorpii
Comments: This small globular is as difficult as M4 is easy. Visible in a 7 x 42 only under the best of conditions. Even a giant reveals little more than a ball of fuzz. A nice challenge.

NGC 6231
Type: Bright open cluster of 120 stars
Size: 15′
Mag.: 6
Location: Less than 1° north of Zeta (ζ) Scorpii
Comments: Fine cluster in any instrument. Small, but very bright. Don't overlook because of its low position in the sky.

COLLINDER 318
Type: Large open cluster of 200 stars
Size: 40′
Mag.: 8.5
Location: In the same field as NGC 6231; 1½° north of Zeta (ζ) Scorpii
Comments: Almost three times larger than nearby NGC 6231, but much fainter and certainly less impressive. Does, however, make an interesting contrast.

SCUTUM

M11, WILD DUCK CLUSTER

Type: Very rich open cluster of hundreds of stars
Size: 14′
Mag.: 5.8
Location: 3½° west and a little south of Lambda (λ) Aquilae
Comments: So rich that it is easily mistaken for a globular. Too compact to be much resolved in a binocular, but a magnificent object. One of the best open clusters.

M26

Type: Open cluster of 30 stars
Size: 15′
Mag.: 8
Location: 6½° southwest of Lambda (λ) Aquilae; less than 1° southeast of Delta (δ) Scuti
Comments: Not as spectacular as M11, but a fine cluster. Small, faint patch of light in a 7 x 42. Speckled cloud of stars in an 11 x 80.

NGC 6712

Type: Globular cluster
Size: 7.2′
Mag.: 8.2
Location: 4° south and west of Lambda (λ) Aquilae; 2° east and a little north of M26
Comments: Barely visible in a 7 x 42 on a good night as a small, fuzzy ball of light. Can be found in the same field as M11 and M26.

SERPENS

M5
Type: Rich globular cluster
Size: 13′
Mag.: 6.2
Location: 7½° southwest of Alpha (α) Serpentis
Comments: One of the best globulars in the sky. Compares favorably with M13. A beautiful ball of light in a 7 x 42. An 11 x 80 reveals a grainy structure in the core itself and resolves some stars at the cluster's fringes.

M16
Type: Open cluster of 60 stars surrounded by a faint emission nebula
Size: 25′
Mag.: 6.5
Location: 6° southeast of Nu (ν) Ophiuchi; 2½ north of M17
Comments: The open cluster is a fine sight in any binocular — even a 7 x 42 resolves many stars. The nebula, however, is a difficult object. It can be glimpsed on a very dark night in an 11 x 80 as a faint haze surrounding the cluster.

TAURUS

HYADES
Type: Very large open cluster of about 40 visible stars
Size: 5.5° (not minutes)
Mag.: 0.5
Location: Centered just west of the star Aldebaran, which is not part of the cluster
Comments: A V-shaped cluster visible to the naked eye. Requires the large field of a 7 x 42 or a 7 x 50 to see the whole cluster. Red Aldebaran provides a nice contrast.

PLEIADES
Type: Open cluster of about 100 stars with 6 or 7 visible to the naked eye
Size: 110′
Mag.: 1.2
Location: 12° northwest of Aldebaran
Comments: Classic object for a binocular. Stunningly brilliant cluster. Compact enough to fit in an 11 x 80, which shows many blue-white stars arranged in rows.

NGC 1647
Type: Open cluster of about 200 stars
Size: 45′
Mag.: 6.4
Location: 3° northeast of Aldebaran on a line with Beta (β) Tauri
Comments: Fine cluster, dense and compact. Nicely resolved in an 11 x 80.

NGC 1746
Type: Open cluster of about 50 stars
Size: 42′
Mag.: 6.1
Location: A little over half the distance from Aldebaran to Beta (β) Tauri
Comments: A more irregular and scattered cluster than nearby NGC 1647. Easily resolved in a 7 x 42.

M1, CRAB NEBULA
Type: Supernova remnant of the A.D. 1054 supernova
Size: 8′ × 6′
Mag.: 8.5
Location: 1° northwest of Zeta (ζ) Tauri
Comments: Famous nebula. Small starlike object in a 7 x 42, but an obvious disk in an 11 x 80.

TRIANGULUM

M33
Type: Very large spiral galaxy
Size: 60′ × 40′
Mag.: 6.5
Location: The same distance southeast of Beta (β) Andromedae that M31 is northwest of it; 4° west and a little north of Alpha (α) Trianguli
Comments: Often a difficult object for small binoculars and even telescopes because of its low surface brightness. Can be found with a 7 x 50 on a very dark night as a subtle, ill-defined glow. Usually an easy object for an 11 x 80.

URSA MAJOR

M81, M82
Type: Bright spiral galaxies
Size: 18′×10′ (M81); 8′×3′ (M82)
Mag.: 7.9 (M81); 9.3 (M82)
Location: At the end of a line that runs from Gamma (γ) through Alpha (α) Ursae Majoris and extends about the same distance as the separation between the stars
Comments: These two galaxies are separated by only 38′. Only M81 may be visible as a nebulous smudge in a small binocular on a poor night. An 11 x 80 shows M82 to be a small sliver of light to the north of M81. Nice pair in a 15 x 80.

M101
Type: Large spiral galaxy
Size: 22′×20′
Mag.: 8.2
Location: 6° north-northeast of Eta (η) Ursae Majoris; forms a triangle with Eta and Zeta (ζ) Ursae Majoris
Comments: A smaller version of M33 in Triangulum. Use at least a 7 x 50 on the darkest night to look for this faint glow. A good test of sky transparency.

VIRGO

M104, SOMBRERO GALAXY
Type: Spiral galaxy
Size: 9′×4′
Mag.: 8.3
Location: 5½° north-northeast of Delta (δ) Corvi
Comments: The best galaxy in Virgo for a binocular. Small nebulous smudge in a 7 x 42 on a good night. An obvious halo of light in a 15 x 80 or 20 x 80.

M84, M86
Type: Elliptical galaxies
Size: 5′×4′ (M84); 7.4′×5.5′ (M86)
Mag.: 9.3 (M84); 9.2 (M86)
Location: First of several galaxies or pairs of galaxies in Virgo that can be located about 8° along a line from Beta (β) Leonis to Epsilon (ϵ) Virginis
Comments: Moderately difficult objects. These ellipticals are located only 17′ apart and usually merge into one smudge in a 7 x 50 on a good night. Remember to use averted vision.

M87
Type: Elliptical galaxy
Size: 7.2′×6.8′
Mag.: 8.6
Location: The next galaxy along the Beta (β) Leonis–Epsilon (ϵ) Virginis line; about 9° east of Beta Leonis
Comments: A little easier than the M84-M86 pair. Faint oval smudge in an 11 x 80.

M60
Type: Elliptical galaxy
Size: 7′×6′
Mag.: 8.8
Location: About 4° west of Epsilon (ϵ) Virginis, just north of Rho (ϱ) Virginis
Comments: Easier than most Virgo galaxies because of its proximity to convenient guide stars. You may also be able to see the 10th-magnitude galaxies M58 and M59 on an excellent night.

M49
Type: Elliptical galaxy
Size: 9′×7.5′
Mag.: 8.4
Location: 7° northeast of Delta (δ) Virginis; forms an equilateral triangle with Delta and Epsilon (ϵ) Virginis
Comments: Similar to M104. An obvious galaxy in an 11 x 80.

VULPECULA

M27, DUMBBELL NEBULA
Type: Planetary nebula
Size: $8' \times 5'$
Mag.: 7.5
Location: A little over 3° north of Gamma (γ) Sagittae
Comments: A ghostly object. Seems to float in the sky with stars for a backdrop. A treat in any binocular.

APPENDIX I

CARE AND CLEANING OF A BINOCULAR

A binocular is a rugged instrument that will give many years of service with only a minimum of care. Here are a few suggestions for proper maintenance.

Always use the neckstrap when you observe. Most binoculars are damaged when dropped or bumped against a hard object. Periodically check the points where the strap attaches to the binocular. These can work loose and allow the strap to slip unexpectedly.

Keep your binocular in a case when not in use or at least keep the lenses covered. Dirt and dust are your lenses' worst enemies.

Clean the lenses only when absolutely necessary. This minimizes wear and damage to the soft optical glass, and also to the much softer coatings.

Most manufacturers recommend a gentle brushing with a camel's hair brush (available at camera stores), followed by gentle wiping with a good quality lens paper to remove dust and grime. Some amateurs use kitchen type cling wrap to remove dust. When peeled off a lens surface, it takes almost all the dust and dirt with it.

Spotting can be removed by first moistening with water or your breath, and gently dabbing with lens paper. There are also several lens cleaning solutions on the market designed to be used on multi-coated lenses. *Never use household chemicals on coated lenses.* They can do irreparable damage to the coatings.

Nor should you attempt to take your binocular apart

unless you are a qualified repairperson. It takes special tools and a lot of expertise to clean, repair, and adjust the internal parts of a binocular correctly. If problems develop, it is best to send the binocular directly to the manufacturer for repair.

APPENDIX II

STAR MAPS FOR BINOCULARS

Since most deep-sky objects for binoculars can be located directly from naked-eye stars, a star map plotted to 6.5 magnitude (naked-eye visibility) will usually suffice. You won't need maps plotted to dimmer magnitudes until you start searching for the smallest, most difficult objects. It is more important for a star map to show accurately a sufficient number of deep-sky objects. Three hundred objects should be considered a minimum. Here are some maps that meet that minimum.

Norton's Star Atlas and Reference Handbook, Arthur P. Norton (600 objects)

The Edmund Mag 6 Star Atlas, Dickinson, Costanzo, Chaple (300 objects)

B.A.A. Star Charts, Wil Tirion (300 objects)

For advanced work, most veterans refer to one of the following:

Atlas of the Heavens, Antonin Becvar (out of print)

Sky Atlas 2000.0, Wil Tirion

Uranometria 2000.0, Tirion, Rappaport, Lovi

I particularly recommend *Sky Atlas 2000.0* as a standard reference for all observers. For the advanced observer *Uranometria 2000.0* shows even more detail, but it is probably too much for a beginner.

ACKNOWLEDGMENTS

A first work is rarely a solo effort. This book is no exception.

To my family for all their encouragement and support, not to mention the timely gift of an 11 x 80 binocular, thank you. To my mother for my interest in nature, thank you. To my wife Janet for her fine illustrations and patience with my observing and writing obsessions, thank you. To my sister Ann for her help in editing and preparing the final copy, thank you. To my brother-in-law Sid Spelts for his superb photographs, thank you. To Rhonda and John McCrory for the gracious loan of their computer, thank you. To my good friend Steve Ninegar for his illustrations of Observer A and Observer B, thank you. To Cabela's and Keith Yocum for allowing me to photograph and test many fine binoculars, thank you.

And of course to my children Kelly and Chrissy (and the one on the way) for proving that a fascination with the stars and heavens is a natural and wonderfully human trait, thank you.

JOHN T. KOZAK

ABOUT THE AUTHOR

John T. Kozak has been observing stars and deep-sky objects with a binocular for 18 years. He is a high school science teacher by training and is currently working on a master's degree in education at the University of Nebraska.

At present, Mr. Kozak is employed as a product specialist with Cabela's, Inc., a nationally known and respected retailer of outdoor equipment. One of his areas of expertise is, of course, binoculars and optics.